Long-Term Response of a Forest Watershed Ecosystem

Clearcutting in the Southern Appalachians

Edited by

WAYNE T. SWANK
JACKSON R. WEBSTER

Oxford University Press is a department of the University of Oxford.
It furthers the University's objective of excellence in research, scholarship,
and education by publishing worldwide.

Oxford   New York
Auckland   Cape Town   Dar es Salaam   Hong Kong   Karachi
Kuala Lumpur   Madrid   Melbourne   Mexico City   Nairobi
New Delhi   Shanghai   Taipei   Toronto

With offices in
Argentina   Austria   Brazil   Chile   Czech Republic   France   Greece
Guatemala   Hungary   Italy   Japan   Poland   Portugal   Singapore
South Korea   Switzerland   Thailand   Turkey   Ukraine   Vietnam

Published in the United States of America by
Oxford University Press
198 Madison Avenue, New York, NY 10016

Long-term response of a forest watershed ecosystem : clearcutting in the southern Appalachians /
[edited by] Wayne T. Swank and Jackson R. Webster.
    pages cm
Includes bibliographical references and index.
ISBN 978-0-19-537015-7 (alk. paper)
1. Clearcutting—Environmental aspects—Blue Ridge Mountains. 2. Forest ecology—Blue Ridge
Mountains. I. Swank, Wayne T. II. Webster, Jackson R.
SD387.C58L66   2014
577.3—dc23
2013029709

9 8 7 6 5 4 3 2 1
Printed in the United States of America
on acid-free paper

Contents

Preface

The Coweeta Hydrologic Laboratory, a research facility of the USDA Forest Service, Southern Research Station, was established in 1934. As noted by Eugene P. Odum, Coweeta represents the longest continuous environmental study on any landscape in North America. The first three decades of research at Coweeta entailed the establishment of a hydrologic network, which included 26 small, forested watersheds. Initial studies included a variety of watershed scale vegetation and land-use experiments elucidating the regulatory effects of forest vegetation on the hydrologic cycle. Subsequently, the research emphasis shifted to interdisciplinary studies of hydrologic processes using the experimented watershed as the basic tool of investigation. At about the same time, cooperative ecosystem research between Coweeta and the Institute of Ecology at the University of Georgia was being developed.

The first 30 years of hydrologic research at Coweeta provided a firm base for forest ecosystem studies at the watershed scale. In 1968 a proposal to the National Science Foundation (NSF) submitted by Phil Johnson of the University of Georgia in collaboration with Forest Service scientists was funded to study the productivity and mineral cycling of natural and manipulated watersheds in the Coweeta basin. In 1971, the initial nutrient cycling studies were broadened as Coweeta became part of the Eastern Deciduous Forest Biome of the US International Biological Program (IBP) funded by the NSF. The breadth of interdisciplinary expertise increased, nutrient budget studies expanded to 16 watersheds in the basin, and process studies were greatly expanded.

As the IBP was reaching conclusion, the Coweeta ecosystem research team saw the need to more vigorously examine forest ecosystem structure and function from a nutrient cycling perspective. A proposal to study the effects of perturbation on nutrient circulation was funded by the NSF in 1974. The treatment selected was a commercial, clearcut, cable-logging study on Coweeta Watershed 7 (WS 7). Later, in 1980, the study was incorporated into the newly established Long-Term Ecological Research (LTER) Program. The study included both practical questions related to forest resource management in the southern Appalachians and fundamental hypotheses on hydrologic and ecological processes derived from previous ecosystem research at Coweeta.

We are frequently asked, What is long-term research? We consider the minimum window of investigation for forest ecosystems to include the life span of the forest of interest, which frequently exceeds a generation of scientists and sometimes two or more generations. The primary goal of this book is to provide a comprehensive update for the long-term study on Coweeta WS 7, which thus far spans a period of nearly 40 years.

The first 11 chapters focus on a variety of specific responses and the recovery of forest and stream processes following cutting. The next two chapters provide findings from two additional long-term experimental forests located in the Appalachian Highlands Physiographic Division that have research objectives similar to those at Coweeta. The last chapter provides a synthesis of 30+ years of research on WS 7.

The majority of support for this research came from the NSF and the USDA Forest Service. We thank NSF for numerous grants through the Division of Environmental Biology and the LTER program. The USDA Forest Service, through both research and the National Forest System organizations, provided support essential to establish and to conduct much of the research. Forest-management specialists on the Wayah Ranger District conducted the timber sale layout, appraisals, on the ground administration, and site preparation work following logging. Nantahala National Forest and Region 8 provided the support that enabled the construction of 8 kilometers of new roads needed to access WS 7. Cook Lumber Company of Franklin, North Carolina, provided and operated the cable-logging technology used in this study. Both the University of Georgia and the Coweeta Hydrologic Laboratory Research Work Unit provided substantial institutional support throughout this study.

We dedicate this volume to several different groups, reflecting the essence of what it takes to achieve success in long-term interdisciplinary ecosystem research. Strong, shared scientific leadership among scientists has been a keynote feature of collaborative research at Coweeta. In the early years of the research, D. A. Crossley, Jr. (aka "Dac") served as the principal investigator (PI) and was charged with the overall coordination of the project. He served in an exemplary role for 22 years. Other senior personnel served as co-PIs: Lindsay Boring, Jim Douglass, Katherine Elliott, Brian Kloeppel, Jennifer Knoepp, Judy Meyer, Carl Monk, Mark Riedel, Wayne Swank, Lloyd Swift, Robert Todd, Jim Vose, Jack Waide, Bruce Wallace, and Jack Webster. Judy Meyer and Dave Coleman served as lead PIs for the Coweeta LTER and strongly supported the continuation of the research on WS 7.

We would also like to acknowledge the contributions of more than 20 graduate students who earned advanced degrees based on their original research, which is an integral part of the research reported in this book. Their findings appear in various chapters and in literature cited throughout the book. Many other graduate students from the University of Georgia, Virginia Tech, and other institutions also contributed in many ways to this study.

We also acknowledge all those who contributed to the outstanding safety record associated with research conducted on WS 7, given difficult field conditions created by steep, slippery slopes, poisonous snakes, yellow jackets, weir cleaning, and other hazardous working conditions.

We owe a special thank you to Kathy Flowers for compiling and maintaining book-related information. We also acknowledge the contributions of all our research support staff, who provided the continuity and dedication required for a successful long-term research program. Field and laboratory technicians who worked on the project included Bob McCollum, Patsy Clinton, Jim Buchanan, Neville Buchanan, Neal Renideau, Mark Crawford, Steve Waldroop, Lee Reynolds, James Wiggins, and Charles Marshall. In the analytical lab, Mike McSwain, Richie Beale, Barbara Reynolds, James Deal, Cindi Brown, Carol Harper, and Wilba Curtis provided high-quality sample analysis. Data management by Bryant Cunningham, Marilyn Payne, Julie Moore, Bruce McCoy, Stephanie Laseter, and Jean Swafford was critical to the success of the project. And a project like this couldn't function without the support of the office administration staff: Mary Lou Rollins, Pat Stickney, and Teresa Moss.

Finally, we owe a special thanks to the chapter authors for their patience throughout this long-term project. This book began with an outline in our 1990 LTER proposal, and while we have reached a certain level of synthesis in the current publication, it is clear that the "Book" on WS 7 is still open with regard to continual, measurable changes in structure and function following cutting and also natural disturbances. While some lessons have been learned from past research, we feel there still remain opportunities to gain more insights into this very complex ecosystem.

Wayne T. Swank
Coweeta Hydrologic Laboratory
USDA Forest Service

Jackson R. Webster
Department of Biological Sciences
Virginia Polytechnic Institute and State University

Contributors

Mary Beth Adams, USDA Forest Service, Northern Research Station, Timber & Watershed Laboratory, Parsons, WV 26287

Amey S. Bailey, USDA Forest Service, Hubbard Brook Experimental Forest, Woodstock, NH 03242

E. F. Benfield, Department of Biological Sciences, Virginia Polytechnic Institute and State University, Blacksburg, VA 24061

Lindsay R. Boring, J. W. Jones Ecological Research Center, Newton, GA 39870

John L. Campbell, USDA Forest Service, Forest Sciences Laboratory, Northern Research Station, Durham, NH 03824

D. A. Crossley, Jr., Department of Entomology, University of Georgia, Athens, GA 30602. Current address: P.O. Box 903, Athens, GA 30603

Christopher Eagar, USDA Forest Service, Forest Sciences Laboratory, Northern Research Station, Durham, NH 03824

Katherine J. Elliott, USDA Forest Service, Coweeta Hydrologic Laboratory, Otto, NC 28763

Damon Ely, Department of Biology, Orange County Community College, Middletown, NY 10940

Stephen W. Golladay, J. W. Jones Ecological Research Center, Newton, GA 39870

Bruce L. Haines, Botany Department, University of Georgia, Athens, GA, 30602 (deceased)

Liam Heneghan, Environmental Science Program, DePaul University, Chicago, IL 60614

James W. Hornbeck, USDA Forest Service, Forest Sciences Laboratory, Northern Research Station, Durham, NH 03824

Jennifer D. Knoepp, USDA Forest Service, Coweeta Hydrologic Laboratory, Otto, NC 28763

James N. Kochendenfer, USDA Forest Service, Northern Research Station, Timber & Watershed Laboratory, Parsons, WV 26287

Stephanie N. Laseter, USDA Forest Service, Coweeta Hydrologic Laboratory, Otto, NC 28763

Kim G. Mattson, Ecosystems Northwest, Mount Shasta, CA 96067

Matthew E. McTammany, Department of Biology, Bucknell University, Lewisburg, PA 17837

Judy L. Meyer, Odum School of Ecology, University of Georgia, Athens, GA 30602. Current address: 498 Shoreland Dr., Lopez Island WA 98261

Robert G. Qualls, Department of Natural Resources and Environmental Science, University of Nevada, Reno, NV 89557

Barbara C. Reynolds, Department of Environmental Studies, University of North Carolina at Asheville, Asheville, NC 28804

Alissa Salmore, Artemisia Land Planning and Design, Pocatello, ID 83201

Timothy D. Schowalter, Department of Entomology, Louisiana State University, Baton Rouge, LA 70803

Wayne T. Swank, USDA Forest Service, Coweeta Hydrologic Laboratory, Otto, NC 28763

James M. Vose, USDA Forest Service, Center for Integrated Forest Science and Synthesis, Department of Forestry and Environmental Resources, North Carolina State University, Raleigh, NC 27695

J. Bruce Wallace, Department of Entomology and Odum School of Ecology, University of Georgia, Athens, GA 30602

Jackson R. Webster, Department of Biological Sciences, Virginia Polytechnic Institute and State University, Blacksburg, VA 24061

LONG-TERM RESPONSE OF A FOREST WATERSHED ECOSYSTEM

1

Programmatic Background, Site Description, Experimental Approach and Treatment, and Natural Disturbances

Wayne T. Swank*
Jackson R. Webster

Introduction

This volume is a synthesis of a long-term interdisciplinary study of watershed ecosystem responses to a forest-management disturbance. Specifically, a commercial clearcut cable logging experiment was initiated on Watershed 7 (WS 7) at the Coweeta Hydrologic Laboratory in 1975 to elucidate ecosystem structure and function by testing hypotheses associated with the hydrologic, biogeochemical, and ecological processes of mixed deciduous forests. Practical forest-management objectives were also integral to the research.

The study as originally proposed evolved from earlier collaborative ecosystem research between the United States Department of Agriculture (USDA) Forest Service and University of Georgia Institute of Ecology investigators as part of the Eastern Deciduous Forest Biome of the US International Biological Program (IBP). The IBP research at Coweeta was conducted on watersheds that had been disturbed 7 to 13 years prior to the initiation of nutrient cycling studies. Therefore, pretreatment and early response data in the IBP studies, along with associated data on mechanisms responsible for ecological responses, were lacking.

In contrast, the WS 7 study was based on a consistent conceptional foundation and theoretical structure. There was also a period of pretreatment calibration. Conceptually, ecosystems were viewed as hierarchical biogeochemical systems in which observable macroscopic properties (solute export, streamflow, leaf area index, etc.) of natural ecosystems were related to their stability (Waide 1988). Theoretical constructs were organized around the resistance-resilience model of

* Corresponding author: Coweeta Hydrologic Laboratory, USDA Forest Service, 3160 Coweeta Lab Road, Otto, NC 28763 USA

ecosystem relative stability across scales of time and space and response to disturbance (Webster et al. 1975; revised by Waide 1988). This model guided both terrestrial research (Monk et al. 1977) and stream research (Webster et al. 1983) at Coweeta, as well as the evaluation of long-term forest responses to intensive management (Waide and Swank 1976; Swank and Waide 1980). A synthesis and thorough discussion of ecological theory derived from the WS 7 experiment is provided by Webster et al. in chapter 14 of this volume.

During the initial 5-year period (1980–1985) of the Long-Term Ecological Research (LTER) Program at Coweeta, the studies on WS 7 provided the centerpiece for ecosystem investigation at the site. Some components of this study have continued up to the present, but new projects were also initiated to address critical gaps in knowledge relative to the structure and function of southern Appalachian ecosystems. The research was then organized according to the 5 core research areas common to all LTER sites: (1) pattern and control of primary productivity, (2) spatial and temporal distribution of populations representing trophic structure, (3) patterns and control of organic matter accumulation in surface layers and sediments, (4) inorganic inputs and movement of nutrients, and (5) pattern and frequency of disturbance to the research site.

Research activity on WS 7 and its reference watersheds was most intense in the first 12 years following harvest, which established the early, rapid changes in ecosystem structure and function. Over the next 15 years, remeasurements of important biogeochemical processes and ecosystem attributes were conducted, along with new studies. Stream research has included long-term studies on the recovery of benthic invertebrates, particulate organic matter, and dissolved organic matter, along with research on leaf breakdown rates, seston transport rates, and nutrient uptake. Long-term components of terrestrial research include continuing measurement of precipitation and stream chemistry; measurements needed to quantify forest succession and associated primary productivity and nutrient dynamics; measurement of litter-soil organic matter and nutrient pools, along with studies on symbiotic nitrogen fixation, log decomposition; recovery of canopy arthropods; and recovery of soil macroarthropod communities and leaf litter decomposition.

The research was designed to address important forest-management issues in the region; in fact, National Forest systems played an important role in the timber sale layout and administration, road construction, and site preparation. When the WS 7 study was initiated, even-aged management (clearcutting) was a primary silvicultural method for regenerating hardwood forests in the southern Appalachians. The conventional method of harvest utilized tractor skidding, with a dense network of roads on steep slopes. An alternative harvest method was cable logging, but the logistics, economic feasibility, soil disturbance, and erosion factors and impacts on water quality were unknown for this extraction method. Research was conducted to address these unknowns and included (1) detailed economic analysis of direct costs per unit of timber volume for cable logging compared with an estimate of the unit costs of conventional logging on the same area (Robinson and Fisher 1982); (2) an intensive study of the source, amount, and fate of sediment associated with 2.95 km of newly constructed access roads (Swift 1988); (3) an evaluation of soil disturbance and erosion from cable logging; (4) a test of a regional empirical model for predicting long-term water yield responses to clearcutting (Swank et al. 2001); and (5) an assessment of harvesting effects on stream water quality. The effects of the management prescription on water, soil, and vegetation sustainability and health are described in individual chapters and synthesized by Webster et al. in chapter 14 of this volume.

Site Description

The study area is located within the Coweeta Hydrologic Laboratory, a 2,185-ha experimental area located in the Nantahala Mountain Range of western North Carolina within the Blue Ridge Physiographic Province, latitude 35°/03' N, longitude 83°/25' W (figure 1.1). The Coweeta climate is classed as Marine, Humid Temperate, with cool summers, mild winters, and abundant rainfall in all seasons (Swift et al.1988). Average annual precipitation varies from 1,700 mm at low elevations (680 mm) to 2,500 mm on upper slopes (> 1,400 mm). Snow usually comprises less than 5% of the precipitation. The underlying bedrock is the Coweeta Group, consisting of quartz diorite gneiss, meta sandstone and peltic schist, and quartzose meta sandstone (Hatcher 1979, 1988). The deeply weathered regolith of the Coweeta basin averages about 7 m in depth.

WS 7, the focus of this book, is a 59-ha, south-facing catchment drained by a second-order stream (Big Hurricane Branch; figure 1.2). Hydrologic measurements began on WS 7 in 1936, shortly after the establishment of Coweeta. A summary of some of the physical and hydrologic characteristics of the WS 7 catchment and also of WS 2, the catchment adjacent to WS 7, which serves as the experimental undistributed reference for assessing many ecosystem responses to disturbance on WS 7, is given in table 1.1. WS 14 (Hugh White Creek) was the primary reference

Figure 1.1   Aerial view of Coweeta basin in 1962 (USDA Forest Service photo); arrow indicates the location of WS 7.

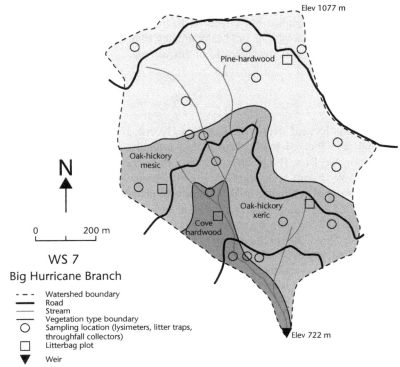

Figure 1.2   Map of WS 7 and Big Hurricane Branch showing roads, streams, major sampling locations, and vegetation types.

watershed used to assess stream biological responses; physical characteristics of this watershed are given by Webster et al. in chapter 10 of this volume. The only management disturbance on WS 7 since the Forest Service's acquisition of the land in 1924, and prior to logging in 1977, was a woodland grazing experiment. From 1941 to 1949, 6 cattle were grazed intermittently over the lower portion of the catchment, duplicating typical forest-use practices of that period. Impacts were limited to soil compaction and overgrazing in the cove forest community but were short-lived (Johnson 1952; Williams 1954). There were no measurable effects of grazing on flow characteristics or stream chemistry 25 years after the termination of the experiment (Swank and Douglass 1977).

Treatment Description

Management prescriptions for the watershed followed National Forest guidelines to achieve desired future conditions for the forested land, and management was applied in three phases: (1) road construction and stabilization, (2) tree felling and logging, and (3) site preparation. Between April and June 1976, three roads

Table 1.1 Physical and hydrologic[a] characteristics of WS 7 and WS 2, Coweeta Hydrologic Laboratory, Otto, North Carolina.

	WS 7	WS 2
Area (ha)	59.5	12.3
Slope (%)	57	60
Aspect	South facing	South-southeast facing
Elevation range (m)	724–1060	716–991
Main stream channel length (m)	1225	480
Soils	Typic Hapludult and Typic Dystrochrept	Typic Hapludult and Typic Dystrochrept
Mean annual precipitation (cm)	189	181
Mean annual flow (cm)	106	99
Range in annual flow (cm)	76–149	59–130
Range in mean daily discharge (L $s^{-1}km^{-1}$)	5–247	5–220
Mean annual quickflow volume (cm)	6.8	8.1
Mean annual evapotranspiration (cm)	83	82

[a] Based on the May-April water year, from 1966 through 1976

with a total length of 2.95 km were constructed for logging access (figure 1.2). Road construction activities incorporated Best Management Practices and some new features of road design standards were applied and evaluated for effectiveness in reducing erosion and sediment movement (Swift 1988). Roadbeds were 4.5 m wide and drained by outsloping (no inside ditches) and broad-based dips. Metal-pipe culverts were installed at three crossings on the perennial stream. Grass was seeded and commercial 10-10-10 fertilizer (N-P-K), and lime was applied on cut-and-fill slopes by a hydroseeder immediately after construction. All roads were seeded by mid-May 1976; but in the last two weeks of May, record storms (38 cm rainfall) eroded both unstable soil and some hydroseeded materials from the roads. Subsequently, the damage was repaired and some road sections were reseeded.

Timber cutting and yarding with a mobile, high-lead cable-yarding system (figure 1.3) began in January 1977 and was completed the following June (figure 1.4). The cable system yarded logs up to 250 m from roads and could suspend logs completely above the ground. Tractor skidding was used on about 9 ha, where slopes were typically less than 20%. The total sale volume was about 2,322 m^3 and this was distributed over 41 ha of the catchment. Due to insufficient volumes of marketable timber, 16 ha on upper slopes and ridges were cut but all wood was left on the ground. The relationship of newly constructed roads to vegetation types, terrestrial sampling sites, and the weir are shown in figure 1.2. Site preparation was completed in October 1977 and this operation consisted of cutting all stems remaining after logging to encourage regeneration. Following logging, most large woody material was removed from the stream channel in accordance with practices used at that time. At the conclusion of the timber sale, roadbeds were reshaped and a light application of grass and fertilizer was applied to disturbed roadbeds and ungraveled sections of the road.

Figure 1.3   View of the high-level cable yarding system on Coweeta WS 7. (USDA Forest Service photo)

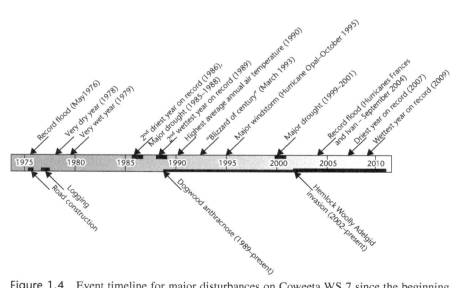

Figure 1.4   Event timeline for major disturbances on Coweeta WS 7 since the beginning of this study.

Table 1.2   Management activities and exposed mineral soil (disturbance) on WS 7, Coweeta Hydrologic Laboratory, Otto, North Carolina.

Activity	Total area of activity (ha)	Total area disturbed (ha)	Percentage disturbance
Permanent road	1.5	1.5	100.0
Cable yarded (uphill)	27.8	1.5	5.6
Cable yarded (downhill)	4.9	0.4	8.5
Tractor skidded	8.9	1.0	10.8
Felled—not logged	15.9	0	0

Source: After Swank et al. (1982)

The exposure of mineral soil associated with each prescription activity on the catchment is given in table 1.2, which shows that half of the exposed soil area was from permanent roads. These roads were also the only significant sources of surface soil compaction and overland flow during storms. It is also apparent that cable logging achieved the goal of minimal soil disturbance and erosion on logged areas.

Natural Disturbances

In long-term studies such as WS 7, it is important to recognize and consider temporal variability in natural disturbances in the context of potential influences on ecosystem processes and responses. Studies described in the subsequent chapters of this book frequently extended over a period of 20 or more years.

In the context of the 73 years of record, the experiment on WS 7, which encompasses the last 30 years, has experienced a variety of record and extreme events (figure 1.4). Beginning in May 1976, shortly after road construction on WS 7 was completed and before cutting and logging, a record storm of 38 cm of rainfall occurred (figure 1.5); this summer convective storm produced the highest peakflow rates ever measured on most of the Coweeta watersheds. The first year after cutting (1978), annual precipitation was 139 cm, or 23% below the long-term annual average. However, 1979 was quite wet, with an annual precipitation of 231 cm, or 28% above average.

Two major droughts occurred during the forest regrowth period. The first drought began in 1985, ended in 1988, and included the second driest year on record (1986), with only 124 cm of precipitation. The other 3 years in this period contained annual precipitation deficits of 33 cm to 54 cm, or 18% to 30% below average precipitation. Major reductions in forest growth and mortality occurred in this period due to water stress and associated southern pine beetle infestations. The drought was followed in 1989 by the wettest year on record (234 cm) at Coweeta. The second drought period spanned a 3-year period beginning in 1989; annual precipitation deficits over the period averaged 40 cm, or 22% above average values. As in the drought of 1985, there was significant forest mortality due to southern pine beetle infestations and reductions in growth.

Figure 1.5 James Buchanan (Jim Buck) contemplating sediment collected in WS 7 weir following record storms in May 1976. (USDA Forest Service photo)

The largest snow of the twentieth century in Macon County occurred in March 1993 and consisted of 0.6 to 1.2 m of snow, high winds, and subzero temperatures. Tree breakage was extensive due to the combined effects of snow, ice, and wind. A substantially greater catastrophic event occurred in October 1995 (figure 1.4) when microburst winds associated with Hurricane Opal created extensive forest wind throw and breakage throughout Macon County and other locations in the region. Within the Coweeta basin, more than 5,000 m^3 of down and damaged timber was salvaged from the existing road network alone. The last windstorm in Macon County approaching the magnitude of this event occurred in 1835 (Douglass and Hoover 1988).

More recently, in September 2004, Hurricanes Frances and Ivan delivered a total of 57 cm of precipitation within a 10-day period. These events generated numerous debris avalanches in Macon County, including severe channel scouring on Coweeta Watershed 37, a high-elevation catchment.

Forest diseases and insect infestations are continuing agents of disturbance in the region. The most significant disease has been the chestnut blight fungus (*Cryphonectria parasitica*), which decimated millions of hectares of its host American chestnut, *Castanea dentata* (Marshall) Borkh. Chestnut blight was first noted in the early 1920s at Coweeta, where chestnut comprised about 35% of the forest basal area (Elliott and Swank 2008). By 1930, most chestnut in the Coweeta basin was infected by the blight, and soon thereafter, most of the overstory trees died. Chestnut still sprouts from roots, but trees seldom live beyond the sapling stage because of the blight. A second disease of significance in the basin is dogwood anthracnose, which is caused by the fungus *Discula destructive* Redlin. First observed at Coweeta in 1989, dogwood anthracnose had an 87% average incidence of infestation on dogwood (*Cornus florida* L.) with a high rate of tree mortality. The disease is still present and a problem in the twenty-first century.

The southern pine beetle (*Dendroctonus frontalis* Zimmerman), in combination with drought, has frequently caused significant mortality in the Coweeta basin. The locust stem borer (*Megacyllen robiniae* Forster) has caused significant mortality in black locust (*Robinia pseudoacacia* L.) at Coweeta, particularly in regenerating stands. Infestations of the hemlock woolly adelgid (*Adelges tsugae* Annand) have spread rapidly since 2002, and high rates of eastern hemlock (*Tsuga canadensis* L.) mortality have been observed (Nuckolls et al. 2009; Ford et al. 2012).

Some of the agents of disturbance illustrated in figure 1.4 and described in this chapter have influenced and altered the structure, function, and related processes in the recovery of the WS 7 ecosystem, as is shown in later chapters of this book.

Literature Cited

Douglass, J. E., and M. D. Hoover. 1988. History of Coweeta. Pages 17–31 in *Forestry Hydrology and Ecology at Coweeta*. W. T. Swank and D. A. Crossley, Jr., editors, Springer-Verlag, New York, New York.

Elliott, K. J., and W. T. Swank. 2008. Long-term changes in forest composition and diversity following early logging (1919-1923) and the decline of American chestnut (*Castanea dentata*). *Plant Ecology* 197: 155–172.

Ford, C. R., K. J. Elliott, B. D. Clinton, B. D. Kloeppel, and J. M. Vose. 2012. Forest dynamics following eastern hemlock mortality in the southern Appalachians. *Oikos* 121: 523–536.

Hatcher, R. D. 1979. The Coweeta Group and Coweeta syncline: major features of the North Carolina: Georgia Blue Ridge. *Southeastern Geology* 21: 17–29.

Hatcher, R. D. 1988. Bedrock geology and regional geologic setting of Coweeta Hydrologic Laboratory in the eastern Blue Ridge. Pages 81–92 *in Forest Hydrology and Ecology at Coweeta*. W. T. Swank and D. A. Crossley, Jr., editors. Springer-Verlag, New York, New York.

Johnson, E. A. 1952. Effect of farm woodland grazing on watershed values in the southern Appalachian Mountains. *Journal of Forestry* 50:109–113.

Monk, C. D., D. A. Crossley, R. L. Todd, W. T. Swank, J. B. Waide, and J. R Webster. 1977. An overview of nutrient cycling research at Coweeta Hydrologic Laboratory. Pages 35–50 in *Watershed Research in Eastern North America: A Workshop to Compare Results*. D. L. Correll, editor. Smithsonian, Washington, DC.

Nuckolls, A. E., N. Wurzburger, C. R. Ford, R. L. Hendrick, J. M. Vose, and B. D. Kloeppel. 2009. Hemlock declines rapidly with hemlock woolly adelgid infestation: impacts on the carbon cycle of southern Appalachian forests. *Ecosystems* 12: 179–190.

Robinson, V. L., and E. L. Fisher. 1982. High-lead yarding costs in the southern Appalachians. *Southern Journal of Applied Forestry* 6: 172–176.

Swank, W. T., and J. E. Douglass. 1977. Nutrient budgets for undisturbed and manipulated hardwood forest ecosystems in the mountains of North Carolina. Pages 343–364 in

Watershed Research in Eastern North America: A Workshop to Compare Results. D. L. Correll, editor. Smithsonian, Washington, DC.

Swank, W. T., J. E. Douglass, and G. B. Cunningham. 1982. Changes in water yield and storm hydrographs following commercial clearcutting on a southern Appalachian catchment. Pages 583–594 in *Hydrological Research Basins and Their Use in Water Resource Planning*. Swiss National Hydrological Service Special Publication, Berne, Switzerland.

Swank, W. T., J. M. Vose, and K. J. Elliott. 2001. Long-term hydrologic and water quality responses following commercial clearcutting of mixed hardwoods on a southern Appalachian catchment. *Forest Ecology and Management* 143: 163–178.

Swank, W. T., and J. B. Waide. 1980. Interpretation of nutrient cycling research in a management context: evaluating potential effects of alternative management strategies on site productivity. Pages 137–158 in *Forests: Fresh Perspective from Ecosystem Analysis*. R. W. Waring, editor. Oregon State University Press, Corvallis.

Swift, L. W. 1988. Forest access roads: design, maintenance, and soil loss. Pages 313–324 in *Forest Hydrology and Ecology at Coweeta*. W. T. Swank and D. A. Crossley, Jr., editors. Springer-Verlag, New York, New York.

Swift, L. W., G. B. Cunningham, J. E. Douglass. 1988. Climatology and hydrology. Pages 35–55 in *Forest Hydrology and Ecology at Coweeta*. W. T. Swank and D. A. Crossley, Jr., editors. Springer-Verlag, New York, New York.

Waide, J. B. 1988. Forest ecosystem stability: revision of the resistance-resilience model in relation to observable macroscopic properties of ecosystems. Pages 383–405 in *Forest Hydrology and Ecology at Coweeta*. W. T. Swank and D. A. Crossley, Jr., editors. Springer-Verlag, New York, New York.

Waide, J. B., and W. T. Swank. 1976. Nutrient recycling and the stability of ecosystems: implications for forest management in the Southeastern United States. Pages 404–424 in *America's Renewable Resource Potential—1975: The Turning Point*. Society of American Foresters, Washington, DC.

Williams, J. G. 1954. A study of the effect of grazing upon changes in vegetation on a watershed in the southern Appalachian Mountains. Thesis. Michigan State University, East Lansing.

Webster, J. R., M. E. Gurtz, J. J. Hains, J. L. Meyer, W. T. Swank, J. B. Waide, and J. B. Wallace. 1983. Stability of stream ecosystems. Pages 355–395 in *Stream Ecology*. J. Barnes and G. W. Minshall, editors. Plenum Press, New York, New York.

Webster, J. R., J. B. Waide, and B. C. Patten. 1975. Nutrient cycling and the stability of ecosystems. Pages 1–27 in *Mineral Cycling in Southeastern Ecosystems*. F. G. Howell, J. B. Gentry, and M. H. Smith, editors. Symposium Series Conf-740513. US Energy Research and Development Administration, Washington, DC.

2

Successional Forest Dynamics
30 Years Following Clearcutting

Lindsay R. Boring*
Katherine J. Elliott
Wayne T. Swank

Introduction

For the past several decades, clearcuts on experimental watersheds have provided an opportunity to examine how these large-scale forest disturbances influence various ecosystem processes, including stream hydrology (Swank and Helvey 1970; Likens et al. 1977; Swift et al. 1988), soil erosion (Hewlett 1979; Van Lear et al. 1985), nutrient cycling (Johnson and Swank 1973; Bormann et al. 1974; Bormann et al. 1977; Likens et al. 1977; Bormann and Likens 1979; Swank and Caskey 1982; Gholz et al. 1985; Boring et al. 1988; Waide et al. 1988; Reiners 1992), and vegetation diversity and successional patterns (Parker and Swank 1982; Gholz et al. 1985; Leopold et al. 1985; Leopold and Parker 1985; Hornbeck et al. 1987; Boring et al. 1988; Reiners 1992; Gove et al. 1992; Elliott and Swank 1994a; Elliott et al. 1997; Elliott et al. 1998). For the investigation of vegetation diversity and successional patterns in the WS 7 clearcut in the Coweeta basin, inventories were conducted 1, 3, 8, 17, 20, and 30 years after disturbance (Boring 1979; Boring et al. 1981, 1988; Boring and Swank 1986; Elliott et al. 1997; Elliott et al. 2002; see table 2.1). Other, related, nitrogen cycling and productivity studies were conducted in early successional black locust (*Robinia pseudoacacia*) stands on both WS 7 (Boring and Swank 1984) and the old-field successional WS 6 (White et al. 1988; Montagnini et al. 1989; Elliott et al. 1998). These Coweeta studies collectively examined the role of dominant early successional species in forest recovery and ecosystem processes and addressed impacts of disturbance on longer-term species composition and diversity.

Much of this research on recovery of forest community structure and ecosystem function was originally proposed as part of a 3-year integrated ecosystem/watershed study funded by the National Science Foundation (NSF), and only after that time

* Corresponding author: Joseph W. Jones Ecological Research Center, 3988 Jones Center Drive, Newton, GA 31770 USA

Table 2.1   Average abundance (for overstory, based on basal area in m²/ha; for ground flora, based on number of plants/m² in 1952 and biomass in g/m² in all other years) and diversity (H', Shannon index) of woody species (≥ 0.5 m height) and herbaceous + woody (< 0.5 m height) ground flora species for three community types in WS 7, Coweeta basin.

Overstory species	Year	F	G	S	Abundance	H'
Cove hardwoods	1974	12	13	14	23.7	2.52
	1977	20	23	28	4.6	2.64
	1979	20	24	32	7.9	2.73
	1984	23	29	36	13.7	2.75
	1993	21	27	36	24.8	2.57
	2008	17	23	28	35.6	2.18
Mesic, mixed-oak	1974	16	21	26	24.9	2.13
	1977	14	17	20	5.3	2.13
	1979	11	14	19	7.3	2.22
	1984	15	18	22	9.2	2.47
	1993	13	14	22	23.8	1.76
	2008	12	20	26	37.7	2.86
Dry, mixed-oak	1974	13	16	19	27.5	2.28
	1977	13	15	19	6.0	2.41
	1979	15	19	25	9.2	2.37
	1984	20	24	30	16.3	2.49
	1993	19	25	36	20.5	2.35
	2008	13	22	30	32.2	2.33
Ground flora species						
Cove hardwoods	1952	12	17	27	16.5	2.52
	1977	12	17	19	33.3	2.49
	1979	12	20	22	97.8	2.19
	1984	16	19	21	37.6	1.85
	1993	14	19	20	8.0	0.82
	2008	23	30	32	15.3	2.73
Mesic, mixed-oak	1952	23	39	49	11.4	3.14
	1977	7	9	10	20.3	1.55
	1979	12	17	18	84.9	2.04
	1984	8	12	13	20.8	1.73
	1993	12	16	16	2.1	1.32
	2008	23	35	41	71.8	2.02
Dry, mixed-oak	1952	18	42	45	13.2	2.40
	1977	10	15	16	43.0	1.99
	1979	12	22	25	46.9	2.28
	1984	16	21	24	17.5	1.65
	1993	16	25	27	3.7	1.90
	2008	16	23	26	28.1	1.91

Note: F, total number of families; G, total number of genera; and S, total number of species present in each community.
Sources: Elliott et al. 1997; Elliott, unpublished.

was its scope and time frame expanded, through continued funding by the NSF's Long-Term Ecological Research Program and the USDA Forest Service's research on biological diversity. Public debate and opposition to clearcutting in the 1980s and 1990s necessitated that more emphasis be placed upon questions related to the impact of forest harvesting on plant species diversity. Fortuitously, the vegetation data sets for the original study had detailed species-level measurements for herbaceous groundcover, shrubs, and trees, and additional studies had also been conducted on the watershed throughout the long-term history of Coweeta. Those early measurements and later finer-resolution sampling at the species and community scales provided additional insights into the whole ecosystem response to clearcutting and improved our fundamental understanding of how early successional vegetation dynamics influence the longer-term recovery of southern Appalachian forests.

We also understand that prior to experimental clearcutting, the forest ecosystem on WS 7 (figure 2.1) was in a highly dynamic state of change and recovery from both climatic and earlier human influences. To keep the subsequent research in perspective, we therefore suggest that regional climatic disturbances and past land-use history be considered in the baseline characterization of the forest ecosystem.

Past human influences included Native American burning and valley agriculture, followed by settlement-period woodland burning and grazing by cattle and hogs. Later, private and federal ownership included intensive early logging practices and total fire suppression (figure 2.2). In the 1930s chestnut blight (*Endothia parasitica*) greatly altered the forest structure because of the removal of the American chestnut (*Castanea dentata*), the most dominant single species of canopy tree at that time (Day and Monk 1974; Elliott and Swank 2008).

Figure 2.1   Mature, mixed-hardwood forest on WS 7, April 1976. (USDA Forest Service photo)

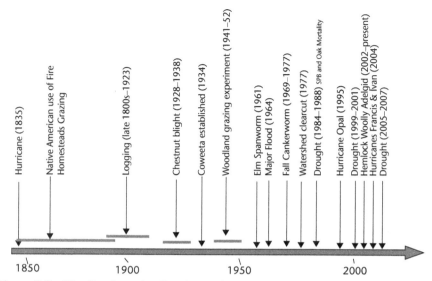

Figure 2.2   Timeline for major disturbances in the Coweeta basin.

Between 1900 and 1923, logging operations occurred over the entire basin, but the cutting was heaviest on the lower slopes, valleys, and accessible coves. Since 1924, human disturbances have been restricted to experimental studies (see Douglass and Hoover 1988 for a complete description of the history of the Coweeta basin). A woodland grazing experiment occurred on WS 7 between 1941 and 1952; six head of cattle were used to assess the impact of woodland grazing on a portion of the watershed. Short-range effects were limited primarily to soil compaction and overgrazing in the riparian area adjacent to the stream (Johnson 1952; Williams 1954).

The individual vegetation and ecosystem research projects on WS 7 have been extensively reported in the literature during the past 30 years. The objective of this chapter is to synthesize the long-term results on successional vegetation dynamics, including species composition, species diversity, biomass and nutrient content, functional roles of key dominant species, and to integrate related short-term studies.

Dynamics of Plant Species Composition and Diversity

Previous papers (Elliott and Swank 1994a; Elliott et al. 1997; Elliott et al. 1998) have described the species-specific changes following clearcutting in three different watersheds in the Coweeta basin. Initially, clearcutting in the southern Appalachians favors shade-intolerant woody species such as *Liriodendron tulipifera* and *R. pseudoacacia*, accompanied by numerous early-successional groundcover species and vines (Boring at al. 1981). After 10–15 years, the shade-tolerant evergreen species *Rhododendron maximum* and *Kalmia latifolia* developed strong dominance in the understory. A substantial reduction in large-seeded and slower-growing species, such as *Quercus* spp., *Carya* spp., *Tilia americana,* and *Aesculus flava,* has resulted from clearcutting, competitive exclusion, and lack of the time and conditions necessary for seed dispersal. These shifts in species composition had significant effects on biomass accumulation and, ultimately, aboveground nutrient accumulation. Although studies have shown a general expansion in evergreen shrubs in the last century (Day and Monk 1974; Monk et al. 1985; Lipscomb and Nilsen 1990; Dobbs 1998; Dobbs and Parker 2004; Elliott and Vose 2012) and a reduced importance of *Quercus* species (Phillips and Murdy 1985; Van Lear 1991; Hedman and Van Lear 1995), with the exception of *Q. prinus* and *Q. coccinea* (Elliott and Swank 2008), clearcutting appears to accelerate this phenomenon.

In WS 7, *L. tulipifera* and *R. pseudoacacia* increased in dominance in all three plant communities after cutting. *Robinia pseudoacacia* could be viewed as a traditional pioneer species that facilitates the growth of the next successional sere (Barnes et al. 1998) because it is short-lived due to early mortality associated with the locust stem borer (*Megacyllene robiniae*) and numerous defoliating insects (Boring and Swank 1984). As a legume, it fixes a substantial amount of nitrogen during its 15 –20 years of site dominance (Boring 1982; Boring and Swank 1984). In contrast, *L. tulipifera*, a shade-intolerant, fast-growing species, reaches the canopy quickly yet is very long-lived (Buckner and McCracken 1978; Burns and Honkala 1990). Once established, even on drier sites, it maintains its canopy position even during drought conditions (Clinton et al. 1993; Elliott and Swank 1994b).

The decline of dogwood (*Cornus florida*) from 1984 to 1997 was strongly influenced by dogwood anthracnose (*Discula destructiva*), a serious disease in southern Appalachian forests since about 1985. Although *C. florida* initially increased in relative dominance, 17 years after cutting (1993), it began a decline that is probably attributable to disease (Elliott et al. 1997). Dogwood anthracnose average incidence of *C. florida* infection was 87% for 1990 in the Coweeta basin (Chellemi et al. 1992; Britton 1993). Also, the loss of initially abundant *C. dentata* sprouts early in succession was due to the chestnut blight that has long been present at Coweeta (Day et al. 1988; Elliott and Swank 2008).

Woody Species Responses

The response of plant communities to clearcutting varied within the watershed. Woody-species richness increased in the cove-hardwood and dry, mixed-oak communities immediately after clearcutting and through 30 years of succession, but remained relatively constant in the mesic, mixed-oak community (table 2.1). In all three communities, there was a trend toward increased diversity (Shannon diversity index, H') in the first 8 years after clearcutting, but none of these differences among years were significant (Elliott et al. 1997). Only in the mesic, mixed-oak community was there a significant decline in diversity, between 1984 and 1993, and then diversity increased again by 2008 (table 2.1). In addition, the number of genera and families also increased in the cove hardwoods and dry, mixed-oak (table 2.1). *Liriodendron tulipifera* increased in dominance in all three communities after clearcutting. *Rhododendron maximum* increased in the cove hardwoods;

R. pseudoacacia increased in the cove hardwoods and mesic, mixed-oak; and *K. latifolia* and *Acer rubrum* increased in the dry, mixed-oak. *Carya* spp. and *Quercus rubra* declined in dominance in the cove hardwoods; *Quercus velutina* and *Carya* spp. declined in the mesic, mixed-oak; and *Pinus rigida* and *Q. velutina* declined in the dry, mixed-oak. In contrast, *R. pseudoacacia* continued to increase in early stages. *Quercus rubra* also decreased in cove-hardwood plots, and *A. rubrum*, important 2 years after cutting, had returned to its precut importance in the community (figure 2.3).

These trends in diversity are similar to those found in other southern Appalachian hardwood forests (Beck and Hooper 1986; Phillips and Shure 1990; Elliott and

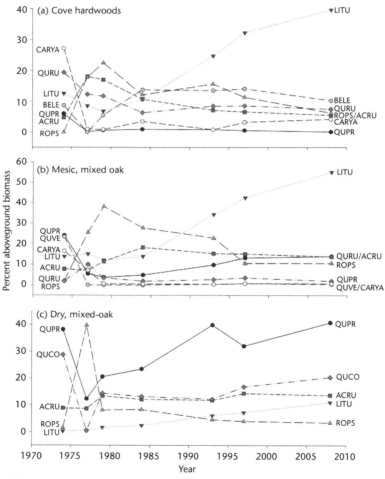

Figure 2.3 Percent aboveground biomass for the three community types in WS 7 through time. Species codes are ACRU, *Acer rubrum*; BELE, *Betula lenta*; CARYA, *Carya* spp.; LITU, *Liriodendron tulipifera*; QUCO, *Quercus coccinea*; QUPR, *Quercus prinus*; QURU, *Quercus rubra*; QUVE, *Quercus velutina*; ROPS, *Robinia pseudoacacia*.

Swank 1994a; Shure et al. 2006) but are somewhat different than those found in northeastern deciduous forests (Gove et al. 1992; Reiners 1992; Wang and Nyland 1993). For example, in the northeast, Gove et al. (1992) showed a decline in tree diversity 10 years after clearcutting, while Reiners (1992) found a gradual decline in diversity and an increase in richness after clearcutting and applying herbicide. In the southern Appalachians, others have found that species composition changed after clearcutting of the mixed-hardwood sites (Beck and Hooper 1986; Phillips and Shure 1990; Elliott and Swank 1994a; Elliott et al. 1998; Shure et al. 2006). These studies also reported that species such as *L. tulipifera*, *R. pseudoacacia*, *C. florida*, and *A. rubrum* increased in dominance after clearcutting; whereas, *Q. rubra*, *Q. velutina*, and *Carya* spp. declined in relative abundance.

Because many species reproduce or sprout and grow rapidly, hardwood forests in the southern Appalachians quickly revegetate and accumulate biomass after disturbance (figure 2.4). Even though revegetation was relatively rapid in WS 7, the successional composition of the forest dramatically changed. For example, *Carya* spp., the leading dominants in the cove-hardwoods community and the third leading dominants in the mesic, mixed-oak community before harvest, comprised less than 3% of the total aboveground live biomass in these communities in 2008 (figure 2.2). Mid- and later successional species, such as *Carya* spp., will probably not become a significant component of the stand for many decades because they disperse seed and grow slowly. Meanwhile, opportunistic species such as *L. tulipifera*, *R. pseudoacacia*, and *A. rubrum* have increased. Because *L. tulipifera* and *R. pseudoacacia* sprout quickly and grow faster than other species, they attain early dominance. *Acer rubrum*, though a shade tolerant species, produced a high number of seedlings in the first few years after cutting in all community types (Elliott et al. 1997). In addition, *A. rubrum* was one of the most prolific sprouting species. Its

Figure 2.4 Regrowth vegetation on WS 7 in June 1979, two growing seasons after harvest. (USDA Forest Service photo)

establishment by both sexual and asexual reproduction contributes to its successful regeneration following disturbance.

The revegetation process on WS 7 was similar to that of other eastern hardwood forests, where sprouts, root suckers, and early successional herbs and vines dominate vegetation after clearcutting (Brown 1974; Ross et al. 1986; Phillips and Shure 1990; Crow et al. 1991; White 1991). In the first year after clearcutting, seedling reproduction and sprout reproduction were about equal, except in the dry, mixed-oak, where sprout reproduction was higher (Boring et al. 1981; Elliott et al. 1997). By 1979, the proportion of stems originating from sprouts increased in all communities. In the dry, mixed-oak community, the high percentage of stems originating from sprouts (81%) probably occurred because seed propagules were scarce and because the xeric forest floor microclimate along the southwest facing slopes and ridges (Swank and Vose 1988) produced a high seedling-mortality rate.

Early and copious production of light, wind-dispersed seeds is generally correlated with the ability to respond to large disturbances (Canham and Marks 1985). Two species that reproduced primarily by seed, *L. tulipifera* and *Q. velutina*, in the cove-hardwoods and mesic, mixed-oak communities provide a striking contrast (figure 2.2). *Q. velutina*, a leading dominant in the mesic, mixed-oak before clearcutting, reproduced only from seed germination or advance seedling growth. *Q. velutina* seedlings totaled 300–700 seedlings/ha, with many present before harvesting (Elliott et al. 1997). Although stumps of *Q. velutina* sprout less frequently than *Q. rubra*, *Q. prinus*, and *Q. coccinea*, the majority of the reproduction after harvest is usually from stump sprouting (Burns and Honkala 1990). In our study, the low basal area for this species after disturbance may be the result of a combination of factors, including low dispersal of seed in the large opening, low survival of seedlings, slow growth of seedlings, and lack of sprouting.

In contrast, *L. tulipifera*, a copious seed producer, established successfully in the cove-hardwood and mesic, mixed-oak communities after clearcutting with 8,000–10,000 seedlings/ha produced during the first year after cutting. In 2008, it was the leading dominant species contributing 55% and 40% to the total aboveground live biomass in the cove hardwoods and mesic, mixed-oak, respectively (figure 2.3). A combination of factors, including prolific seed production, extended seed viability in the forest floor, survival of new germinants, relatively fast growth, tolerance of the codominant *R. pseudoacacia,* and some stump sprouting are responsible for this success. By 1993, *R. pseudoacacia* declined, but its fixed nitrogen apparently enhanced *L. tulipifera* growth in contrast to other site competitors (Apsley 1987).

Ground Flora Responses

Ground flora was in a transitional state between early and late successional species 17 years after clearcutting (Elliott et al. 1997). Early successional *Aster*, *Solidago*, and *Eupatorium* species declined in abundance because woody species grew rapidly and the canopy closed within 3 to 6 years. Late successional herbaceous species had not become abundantly established, which caused a significant decline in ground flora diversity in the cove hardwoods and mesic, mixed-oak. Total number of plant species present (woody + ground flora) increased in all three communities during the first 3 years after cutting. Then, total species remained relatively constant in the cove hardwoods and mesic, mixed-oak from 1979 to 1993; however, total species continued to increase through 1993 in the dry, mixed oak. Thirty years after cutting (2008), total species had reached the precut forest condition in the cove-hardwoods and mesic, mixed-oak communities, yet the dry, mixed-oak community remained lower than the precut forest (table 2.1).

In general, ground flora diversity (H') declined from 1977 to 1993 in the cove hardwoods and mesic, mixed-oak communities but did not decrease significantly in the more open-structured dry, mixed-oak (table 2.1). In every community on the watershed, more species were present before than in the years after clearcutting until 2008 (table 2.1). This pattern parallels results reported by Gove et al. (1992), where diversity of all plant species (overstory and ground flora combined) decreased 10 years after clearcutting in New Hampshire. Nixon and Brooks (1991) found that herbaceous species diversity peaked in year 3 after clearcutting a deciduous forest in east Texas then subsided through year 9. Similarly, in a chronosequence study of diversity trends following clearcutting (6 stands ranging from 1 to 26 years old) in Allegheny hardwoods, Yorks and Dabydeen (1999) found no significant differences between diversity indices and stand age, but they did show a trend in higher herbaceous diversity in recent clearcuts (4 years old) than control sites (> 75 years old). In addition, Gilliam et al. (1995) and Gilliam (2002) compared two 20-year-old clearcut forests and two mature forests (> 70 years old) in the Allegheny Mountains of West Virginia. They reported no significant differences between young (20-year-old clearcut) forests and mature forests in ground flora diversity. In northern Wisconsin, Brosofske et al. (2001) also found no differences in diversity (H') between young (10–15 years old) and mature northern hardwood forests, but they did find higher richness and diversity in clearcut (4–8 years old) forests compared to the young and mature forests. In WS 7, it was 30 years after clearcutting when species richness and diversity of the cove hardwoods exceeded and the mesic, mixed oak reached the precut forest condition (table 2.1).

In WS 7, the abundance (i.e., g biomass/m²) of ground flora was also lower in 1993 compared to 1984; 79% less in the cove hardwoods; 90% less in the mesic, mixed-oak; and 79% less in the dry, mixed oak. With the growth of overstory trees and canopy closure, the number of early successional, shade-intolerant genera, such as *Erechtites, Solidago, Eupatorium, Panicum, Rubus,* and *Aster* had declined. Late successional, shade-tolerant species such as *Viola, Galium, Sanguinaria, Uvularia, Veratrum,* and ferns had not become well established in the watershed. WS 7 was still in a transition state between early and late successional species abundance. After 30 years (2008), biomass of ground flora increased substantially because late successional species were again abundant. For example, in the mesic, mixed-oak, two species contributed 63% to the total ground flora biomass, *Polystichum acrostichoides* and *Solidago curtissii,* a common forest herb. Other spring ephemerals may have recovered over time, however, the timing of measurements prevented examining the response of these species, such as *Trillium, Anemone,* and *Claytonia.* Because spring ephemerals respond to changes in temperature and light (Collins et al. 1985), clearcutting may have reduced these species via heat stress or triggered changes in seasonal phenology, growth, and reproductive potential in the earlier years after cutting.

After 30 years of recovery, ground flora species diversity and richness in WS 7 was similar to a nearby 30-year-old clearcut watershed (WS 13) with the same community types (Elliott and Swank 1994a). Time since disturbance seems to be an overriding factor even though there are other differences between these two watersheds including: the larger spatial scale of disturbance in WS 7 (59 ha cut in WS 7 vs. 16 ha cut in WS 13); southwest-facing aspect of WS 7, which receives higher solar radiation than the east-facing aspect of WS 13.

Changes in Forest Species Composition

Clearcutting favors shade-intolerant woody pioneering species, such as *L. tulipifera* and *R. pseudoacacia*, and shade-tolerant woody understory species, such as *R. maximum* and *K. latifolia*. There are strong positive responses to clearcutting by these two markedly different groups of plants. This strongly indicates that retention of formerly dominant and ecologically critical mid- and later successional overstory species (especially *Quercus* and *Carya*) is questionable. Their regeneration with successional development is dependent upon past ecological conditions, which included a relatively open understory light environment without evergreen shrub dominance and periodic presence of fire (Tainter et al. 1984; Van Lear 1991).

Both anthropogenic (e.g., chestnut blight, fire exclusion, and cattle grazing) and natural disturbances (e.g., drought) shaped forest composition in WS 7 before clearcutting. The composition of southern Appalachian forests has been significantly altered by the loss of *C. dentata* (Woods and Shanks 1959; Arends 1981; Day et al. 1988; Busing 1989). Chestnut blight was a major impact in the Coweeta basin, because *C. dentata* made up an estimated 35%–40% of the basal area of some forest stands (Day et al. 1988). Canopy openings due to *C. dentata* mortality in the late 1930s has favored the expansion of *Rhododendron* (Day and Monk 1974; Monk et al. 1985; Elliott and Vose 2012); and fire suppression has likely favored the expansion of *Kalmia latifolia* on upper slopes and ridges (Phillips and Murdy 1985; Van Lear 1991). *Rhododendron* often dominates understory canopy layers in riparian stands and adversely affects development and richness of herbaceous and understory stratum (Baker 1994; Hedman and Van Lear 1995). Heavy cattle grazing can also have a dramatic effect on species richness and diversity. For example, Williams (1954) found a loss of 31 species in the cove-hardwood community of WS 7 during a 12-year period (1940–1952) of heavy grazing; however, the mesic, mixed-oak and dry, mixed-oak types showed little to no loss of species on slopes and ridges where cattle were less likely to travel. In addition, severe droughts have caused substantial tree mortality in the southern United States (Hursh and Haasis 1931; Tainter et al. 1984; Stringer et al. 1989; Starkey et al. 1989; Smith 1991; Clinton et al. 1993; Elliott and Swank 1994b).

Although separating the cumulative effects on vegetation dynamics is difficult, this complex of disturbances is typical of conditions throughout much of the southern Appalachians. The cumulative vegetation responses to clearcutting and other disturbances found here are indicative of the regional responses of forests since the early twentieth century. Other influences of regional atmospheric pollution and climate change may also have an undefined influence on species richness and community composition.

It should be stressed that the disturbance legacies of the southern Appalachians are not simply cumulative but are probably synergistic. Fire suppression coupled with both early-century exploitive logging and chestnut blight impacts have resulted in great expansion in the distribution and increased importance of fire-intolerant, thin-barked opportunistic species including *A. rubrum, L. tulipifera, K. latifolia,* and *R. maximum*. Fire suppression in isolation of drastic canopy disturbance likely would not have changed forest composition so rapidly. Furthermore, structural disturbances with continued low intensity burning would have likely favored regeneration of more *Quercus, Carya,* and *Pinus* species, all moderately fire-tolerant.

Before clearcutting, many of the opportunistic species were well poised to respond aggressively to this disturbance. Because of the series of preceding forest structural disturbances, they had already increased in importance, as documented with *L. tulipifera* on the clearcuts WS 13 and WS 6 (Elliott and Swank 1994a; Elliott et al. 1998), and by the other species responses to clearcutting on WS 7 (Elliott et al. 1997). Their life-history characteristics and physiological adaptations made them highly adaptable to take advantage of these disturbance-compounded forest conditions.

Dynamics of Forest Biomass, Leaf Area, and Nutrient Accumulation

Early Successional Recovery

Rapid regeneration of foliar biomass and leaf area index (LAI) are important in the recovery of biotic control of ecological processes such as net primary production (NPP), evapotranspiration, and retention of nutrient and organic capital (Waring and Schlesinger 1985; Mann et al. 1988; Crowell and Freedman 1994; Bolstad et al. 2001; Swank et al. 2001). Sprouting and rapid vegetative growth provide mechanisms for rapid recovery in eastern deciduous forests. In WS 7, leaf biomass and LAI recovered within 8 years after clearcutting (figure 2.5), similar to findings in other clearcut successional forests at Coweeta (Swank and Helvey 1970; Swift and Swank 1981). This rapid recovery of LAI increased evapotranspiration, which reduced the extra water yield observed after clearcutting; by the seventh year, annual streamflow had returned to pretreatment levels (Swank et al. 2001). In addition, the high rate of NPP and sequestration and storage of nutrients during early succession substantially moderated the loss of solutes from the watershed (Swank et al. 2001). At age 30 years, the aboveground biomass on WS 7 had exceeded the precut biomass (table 2.2). Similarly, the accumulations of nutrients aboveground showed major recovery in two decades ranging from 57% to 83% for P and 35% to 50% for Mg, depending on community type. Substantial spatial variability in nutrient accumulation was present across the three communities, reflecting soils and species nutrient requirements.

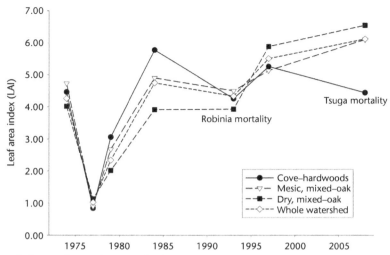

Figure 2.5   Leaf area index (m²/m²) for the cove hardwoods (closed circles), mesic, mixed-oak (*open triangle*), and the dry, mixed-oak (*closed square*) communities in WS 7 and the whole watershed (*open diamond*).

In the first 1–3 years following cutting, herbaceous-layer species including *Rubus*, *Panicum*, *Aster*, and *Solidago* spp. contributed significantly to accumulation of aboveground biomass, LAI, and nutrient capital (Boring et al. 1981; Boring 1982; Boring and Swank 1986). However, soon after cutting, woody species became increasingly abundant, and proportionately less biomass and nutrient capital were found in herbaceous-layer species (figure 2.6). For example, by 1993, herbaceous-layer biomass contributed less than 0.1% to total aboveground biomass in any of the three community types (Elliott et al. 1997). These findings are consistent with other closed canopy forests in the southern Appalachians (Day and Monk 1974; Busing et al. 1993). Although the ground vegetation contributed considerably to aboveground biomass the first few years following disturbance in WS 7 (Elliott et al. 2002), the contribution was proportionately less than found in northern forests (Crow et al. 1991; Reiners 1992; Crowell and Freedman 1994).

Large forest openings significantly change the forest floor microclimate for all residual biota, including woody seedlings and late successional herbaceous species (Phillips and Shure 1990). Other investigators at Coweeta have found that clearcutting on WS 7 increased mean monthly temperatures at the litter/soil boundary for the period May–October by 8°C to 11°C the first year after cutting, reduced forest floor litter moisture, increased soil moisture (Swank and Vose 1988), altered microarthropod activity in the litter (Seastedt and Crossley 1981; Seastedt et. al. 1983a), and reduced first-year decomposition of woody litter, especially on xeric south-facing slopes (Abbott and Crossley 1982). The increase in woody leaf area index by the third year after clearcutting resulted in forest floor shading, amelioration of the altered forest floor microclimate, and dampening of environmental effects on forest floor biota and their processes. Although seedlings and ground

Table 2.2   Aboveground biomass, leaf biomass, and aboveground nutrient mass in three community types and weighted watershed values before cutting (1974) and after cutting (1977–2008).

	Year	Aboveground	Leaf mass	Nitrogen	Phosphorus (kg/ha)	Potassium	Calcium	Magnesium
Cove Hardwoods	1974	204,330	3650	516.9	74.4	503.3	768.0	110.8
	1977	1,054	691	9.0	0.6	5.3	12.8	3.0
	1979	6,299	2434	55.4	4.1	31.9	59.1	12.1
	1984	26,263	3964	133.7	12.6	97.6	120.5	23.5
	1993	67,454	3135	203.2	30.8	166.6	200.4	33.6
	1997	81,692	3236	251.9	42.7	223.2	237.8	38.8
	2008	183,159	4669	—	—	—	—	—
Mesic, mixed-oak	1974	184,973	3354	482.3	70.5	415.6	643.9	99.0
	1977	1,255	705	14.7	1.0	8.5	14.1	3.0
	1979	6,779	2562	72.5	4.7	40.0	67.8	14.2
	1984	27,896	3883	150.8	13.6	108.8	118.0	23.4
	1993	77,720	2684	262.2	42.1	207.7	212.1	36.8
	1997	96,617	2936	318.0	58.6	290.6	260.4	43.2
	2008	212,835	4239	—	—	—	—	—
Dry, mixed-oak	1974	127,196	3467	339.8	42.0	262.7	462.3	70.4
	1977	1,562	928	16.7	1.1	8.8	14.2	3.5
	1979	3,987	1701	37.0	2.8	21.7	30.0	8.1
	1984	22,851	3163	114.4	10.9	79.2	90.7	17.8
	1993	58,705	3332	143.5	17.1	105.1	169.4	26.3
	1997	85,429	3900	216.7	27.4	163.9	245.4	36.0
	2008	152,449	5169	—	—	—	—	—
Watershed	1974	153,203	3466	401.7	53.9	336.9	553.6	83.6
	1977	1,406	835	15.1	1.0	8.1	13.9	3.2
	1979	5,034	2022	48.6	3.4	27.8	43.9	10.2
	1984	24,627	3462	126.4	11.9	89.3	102.0	20.0
	1993	64,763	3140	182.2	25.4	140.0	184.7	30.0
	1997	87,661	3559	247.3	37.4	204.4	248.0	38.3
	2008	172,152	4861	—	—	—	—	—

Note: Plant tissue for nutrient analysis was not collected in 2008.

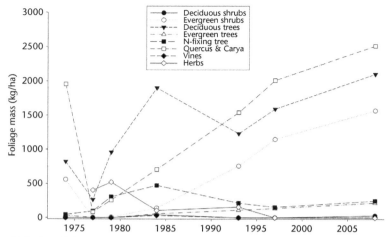

Figure 2.6 Leaf biomass of different growth forms for the whole WS 7.

flora may have been impacted by high mortality immediately after clearcutting, canopy closure within three years enabled a subsequent rapid recovery of structural and functional forest processes.

Spatial Variability of LAI

In the mixed hardwood forests within the Coweeta basin, variation in annual net primary production (ANPP) and LAI have been attributed to an elevation gradient (with higher precipitation and lower temperatures at high elevations) and a topographic/terrain gradient (with higher temperatures and lower soil moisture on ridges) (Bolstad et al. 2001). This elevation gradient and terrain shape index also influences species distributions (Elliott et al. 1999) and therefore affects potential biomass and nutrient accumulation following a major disturbance such as clearcutting.

Our estimates for LAI of the precut (1974) forest of WS 7 (70+ years old) were slightly higher than for a comparable forest type (i.e., mesic, mixed-oak) on a south-facing watershed (Monk et al. 1970), which received higher solar radiation (leading to higher temperatures). Monk et al.'s (1970) value for LAI was similar to that of a young, mixed-deciduous forest in New England (Delucia et al. 1998), but it was less than other reported LAI values within the Coweeta basin. In another undisturbed watershed, where evergreen species contributed up to 35% of the total standing crop of leaves (Monk and Day 1984), LAI was estimated at 6.2 m^2/m^2 (Monk and Day 1988). In mature (75+ year old) forests within the Coweeta basin, Bolstad et al. (2001) reported a range in LAI from 2.7 to 8.2 m^2/m^2, depending on elevation and terrain position (ridge, sideslope, or cove), with LAI decreasing as elevation increased (a function of lower temperatures) and as terrain position increased (ridges having the highest terrain shape index; a function of lower soil moisture (Helvey and Patric 1988; Yeakley et al.

1998). For mature forests within 800–1000 m elevation (the elevation range of WS 7), LAI ranged from 5.0 to 6.5 m^2/m^2 (Bolstad et al. 2001). However, we did not find a decrease in LAI from cove to ridge, primarily due to the abundance of *K. latifolia* and *R. maximum* that contributed substantially to total LAI in the dry, mixed-oak community located on upper slopes to ridges. Across communities, LAI in the 30-year-old forest of WS 7 was already greater than the LAI in a 50-year-old forest reported by Monk et al. (1970) and nearly as high as the 75+-year-old mature forests at similar elevations reported by Bolstad et al. (2001).

Ecosystem Influences of N_2 fixation

Later in succession (10–15 years), though forest structure was still partially dominated by the N_2-fixing tree *R. pseudoacacia*, WS 7 NO_3-N export increased above previously observed levels, indicative of significant changes in ecosystem N-cycling (Swank et al. 2001). Forest composition was changing because of intense stand competition and insect attacks by defoliators and stem borers upon *R. pseudoacacia* (Boring and Swank 1984; Apsley 1987). The death and defoliation of so many of these trees resulted in elevated organic and inorganic nitrogen flux.

The successional *R. pseudoacacia* stands were estimated to have fixed 30–75 kg N ha^{-1} yr^{-1}. Related N-cycling studies of senescent, defoliated stands of *R. pseudoacacia* in an old-field successional forest documented: (a) high leaf litter N transfers (60 vs. 21 kg N $ha^{-1}yr^{-1}$ for similar pine hardwoods) with moderate decomposition rates (White 1986; White et al. 1988); (b) high root litter transfers, high substrate quality and decomposition rates (Grimm 1988); (c) elevated growing season N deposition (> 30 vs. 8 kg N ha^{-1} yr^{-1} for pine hardwoods) via insect frass and fine particulates associated with canopy defoliation (Boring et al. 1987; Seastedt et al. 1993b); (d) elevated net throughfall flux of organic N and NH_4-N (8 vs. < 1 kg N ha^{-1} yr^{-1} for pine hardwoods; Boring et al. 1987); and (e) elevated soil net mineralization and nitrification rates (Montagnini et al. 1986; Montagnini et al. 1989). This N_2-fixing species plays a significant role in the regeneration phase of southern Appalachian forests following disturbance. Thus, *R. pseudoacacia*'s influence is great in early sequestration of nutrients, moderation of microclimatic, elevation of N inputs, availability, and flux, and its legacy of residual N to the later successional forest.

Regional Comparisons

Accumulation of aboveground biomass and nutrients in live vegetation was faster in the southern Appalachian hardwood forest on WS 7 than in northeastern hardwood forests. Reiners (1992) found that biomass accumulation was 52 Mg/ha (38% of the total for a nearby, mature watershed) 20 years after clearcutting a hardwood forest at Hubbard Brook, New Hampshire. In WS 7, biomass accumulation was much higher (56%–85%) than in the clearcut watershed in Reiners's (1992) study. In addition, the aboveground pools of nutrients were almost twofold higher in WS 7 than in the northern hardwood forest at Hubbard Brook. However, in the Hubbard Brook study, the experimental treatment differed from commercial clearcutting in that timber was not removed and regrowth was suppressed with herbicides for the first 3 years following cutting. One of the effects of herbicide treatment may have been to decrease regeneration by sprouting, which is relatively important in northern hardwood forests (Martin and Hornbeck 1989; Reiners 1992). In a northern hardwood forest in Nova Scotia, Crowell and Freedman (1994) reported lower aboveground biomass and nutrient stocks for woody species 8 and 20 years after clearcutting than those reported for WS 7 (Elliott et al. 2002).

Several factors explain higher biomass and nutrient accumulation in the southern Appalachian hardwood forests, when compared to the northern hardwood forests of New England (Reiners 1992; Crowell and Freedman 1994): the pre-humid climate (higher temperature and precipitation), prolific sprouting ability of hardwoods (Kays and Canham 1992; Elliott et al. 1997), and abundance of the N-fixing *R. pseudoacacia*. Not only does *R. pseudoacacia* fix a substantial amount of N in young stands (Boring and Swank 1984), it also accumulates large quantities of N in leaves, branches, stems, and roots and facilitates growth of associated woody species (Apsley 1987). Much higher-N standing stocks were measured in 17- and 38-year-old dense stands of *R. pseudoacacia* (Boring and Swank 1984) than found in mixed hardwood stands on WS 7.

Southern Appalachian forests have the potential to accumulate more aboveground biomass and nutrients in all steps of succession than do most North American deciduous forests (O'Neill and DeAngelis 1981; Cannell 1982; Grier et al. 1989; Busing et al. 1993; Reich et al. 1997; Barnes et al. 1998; Waring et al. 1998). For example, in the Great Smoky Mountains of eastern Tennessee, aboveground biomass estimates for cove-hardwood, old-growth forests ($\approx$ 400 year old) was > 300 Mg/ha (table 2.3) with a corresponding ANPP of 6.3 to 8.6 Mg ha^{-1} yr^{-1}. In young stands (42–63 years old), aboveground biomass ranged from 216 to 277 Mg/ha with a corresponding ANPP of 11.7–13.1 Mg ha^{-1} yr^{-1} (table 2.3). By contrast, a northern hardwood forest in central New Hampshire, in the absence of logging (Martin and Bailey 1999) and a 70-year-old maple-oak forest in northern Michigan (Crow et al. 1991) (table 2.3) had less aboveground biomass than what we found in cove-hardwoods or mesic, mixed-oak communities of WS 7 before clearcutting ($\approx$60 year old forest) (table 2.3). Monk et al. (1970) estimated aboveground live biomass of 145 Mg/ha for an oak-hickory forest in the Georgia piedmont, and Day and Monk (1974) estimated aboveground live biomass of 140 Mg/ha in a north-facing watershed in the Coweeta basin (heavily logged between 1900 and 1923). Theirs were less than our estimates of aboveground live biomass in the precut (1974) forest (also heavily logged between 1900 and 1923) for either the cove-hardwood or mesic, mixed-oak communities of WS 7, a south-facing watershed. Within the Coweeta basin, for mature stands (75+ year old), Bolstad et al. (2001) reported a range of ANPP from least productive (5.2 Mg ha^{-1} yr^{-1}) on ridges at high elevation to most productive (11.7 Mg ha^{-1} yr^{-1}) in cove hardwoods. In a 50-year-old *L. tulipifera* stand in Tennessee, Cole and Rapp (1981) reported an aboveground biomass of 125 Mg/ha and nutrient standing stocks of 305 kg N ha^{-1}, 47 kg P ha^{-1}, 173 kg K

Table 2.3   Aboveground biomass of eastern North American deciduous forests.

Location	Forest type	Forest age (years)	Aboveground biomass (Mg/ha)	Citation
Great Smoky Mountains, TN	Cove-hardwood	~400	326–384	Busing et al. 1993
Great Smoky Mountains, TN	Cove-hardwood	42–63	216–277	Busing et al. 1993
White Mountains, NH	Northern hardwood	100	208	Martin and Bailey 1999
Hubbard Brook, NH	Northern hardwood	20	152	Reiners 1992
Hubbard Brook, NH	Northern hardwood	~70	400	Reiners 1992
Northern, MI	Maple-oak	70	151	Crow et al. 1991
Eastern, USA	Hardwoods	< 100	159	Brown et al. 1999
Piedmont, GA	Oak-hickory	50	145	Monk et al. 1970
Southern Appalachians, NC	Pine-hardwood	70	92–184	Vose and Swank 1993
Southern Appalachians, NC	Mixed deciduous	50	140	Day and Monk 1974
Southern Appalachians, TN	*L. tulipifera*	50	125	Cole and Rapp 1981
Mid-Atlantic region	Oak-hickory	< 100	199	Jenkins et al. 2001
Mid-Atlantic region	Maple-beech-birch	< 100	207	Jenkins et al. 2001
Mid-Atlantic region	Oak-pine	< 100	162	Jenkins et al. 2001
White Mountains, NH	Northern hardwood	~80–120	192	Goodale and Aber 2001
White Mountains, NH	Northern hardwood	>150	261	Goodale and Aber 2001
Northern WI	Aspen-maple-birch	~80	92–118	Crow 1978
South-central CN	Oak-hickory-maple	80	166–185	Tritton et al. 1987

Note: Mid-Atlantic included 7 states; NY, PN, NJ, MD, DL, WV, VA and aboveground tree biomass was estimated from FIA data.

ha^{-1}, and 456 kg Ca ha^{-1}. In WS 7, within only ≈20 years of cutting, aboveground biomass was 66%–78% of that reported by Cole and Rapp (1981), depending on community type; nutrient standing stocks were 60%–69% for N, 53%–100% for P, 88%–100% for K, and 52%–57% for Ca.

In nearby mature (≈70 years old), pine-hardwood stands in the southern Appalachians, Vose and Swank (1993) reported a range in aboveground biomass (table 2.3). Although before cutting, the dry, mixed-oak community in WS 7 was within this range of values (table 2.2), it had a larger component of oaks (93.3 Mg/ha), only a minor component of *P. rigida*, and less *K. latifolia* (3.77 Mg/ha). However, with only ≈20 years of regrowth, biomass accumulation of *K. latifolia* was nearly equal to its precut value, and biomass of *R. maximum* was 4.6 times greater than before cutting in the dry, mixed-oak community. In addition, leaf biomass of these two evergreen species was 31% of total leaf biomass. Because these two species can retain foliage for up to 3 years, only 10%–15% of their total leaf nutrient standing crop is lost annually through litterfall, and because sclerophyllous foliage decomposes slowly (Monk et al. 1985), the increased abundance of *K.*

latifolia and *R. maximum* in WS 7 could have long-term implications for ecosystem processes such as decomposition and nutrient cycling.

Long-Term Ecosystem Implications

The exceptionally high species diversity and ecosystem complexity of southern Appalachian forests should be considered when assessing changes in the structure and function of the forest after clearcutting. The high plant-species richness, diverse functional groups, interactions among species, and the variety of life histories result in their diverse light-, water-, carbon-, and nutrient-use strategies. This is reflected in their requirements for unique successional niches. Recent debates have raged in the ecological community about the potential relationships between species richness and the stability of ecosystem function.

A pre-humid climate, prolific hardwood sprouting, an abundance of *R. pseudoacacia,* and the observed changes in species composition have important long-term implications for ecosystem function (Elliott et al. 2002). Clearcutting favored shade-intolerant, fast-growing species, such as *L. tulipifera* and *R. pseudoacacia*; subdominant *A. rubrum*; and shade-tolerant understory shrubs, such as *Rhododendron maximum* and *K. latifolia*. The abundance of large-seeded, slow-growing species, such as *Quercus* spp. and *Carya* spp. declined after clearcutting (Elliott et al. 1997). *Robinia pseudoacacia* and *L. tulipifera* lose their foliage in early fall and have high leaf-nutrient concentrations; *L. tulipifera* and *A. rubrum* foliage decomposes quickly. *Quercus* spp. and *Carya* spp. lose their foliage in late fall through the winter and have lower leaf-nutrient concentrations; and their foliage decomposes slower than many of the other overstory species that replaced their dominance (Cromack 1973; White et al. 1988). These reported shifts in overstory species composition could result in forest floor litter with higher nutrient concentrations, lower C/N ratios, and subsequently faster decomposition rates.

However, the dynamics of the increasingly dominant evergreen understory *K. latifolia* and *R. maximum* may be as important to future ecosystem functions as the changes in the overstory (Elliott et al. 2002). Both species have also increased in abundance in the Coweeta basin since the loss of *C. dentata* from the overstory, and clearcutting on WS 7 further increased their distribution, basal area, and density (Elliott et al. 1997). In areas of the watershed where *K. latifolia* and *R. maximum* are abundant, the contribution of their ericaceous leaves to litterfall may reduce litter quality, even with *L. tulipifera* and *R. pseudoacacia* in the overstory. *Kalmia latifolia* and *R. maximum* retain foliage for several years and have low leaf-nutrient concentrations, and the scherophyllous leaves decompose slowly (White 1986; White et al. 1988). The organic soils and root mats that these ericads form may potentially alter soil pH and decrease forest floor decomposition rates (Knoepp et al. 2000). Because decreased rates of litter decomposition would slow nutrient cycling rates within an ecosystem, the composition and quality of litter have important implications in terms of nutrient loss and retention, soil nutrient availability, and soil quality (Knoepp et al. 2000). Species differences in evapotranspiration would also have long-term effects on hydrologic cycles at the watershed scale

(Swank et al. 2001). These major changes in overstory and understory composition affect forest structure, organic matter quality, and water use; consequently, these vegetation dynamics could result in different watershed hydrologic and biogeochemical responses in the future.

Literature Cited

Abbott, D. T., and D. A. Crossley, Jr. 1982. Woody litter decomposition following clearcutting. *Ecology* 63: 35–42.

Apsley, D. K. 1987. Growth interactions and comparative water relations of *Liriodendron tulipifera* L. and *Robinia pseudoacacia* L. Thesis. University of Georgia, Athens.

Arends, E. 1981. Vegetation patterns a half century following the chestnut blight in the Great Smoky Mountains National Park. Thesis. University of Tennessee, Knoxville.

Baker, T. T. 1994. The influence of *Rhododendron maximum* on species richness in the riparian zone of Wine Spring Creek. Thesis. Clemson University, Clemson, South Carolina.

Barnes, B. V., D. R. Zak, S. R. Denton, and S. H. Spurr. 1998. *Forest Ecology.* 4th ed. John Wiley & Sons, New York, New York.

Beck, D. E., and R. M. Hooper. 1986. Development of a southern Appalachian hardwood stand after clearcutting. *Southern Journal of Applied Forestry* 10: 168–172.

Bolstad, P. V., J. M. Vose, and S. G. McNulty. 2001. Forest productivity, leaf area, and terrain in southern Appalachian deciduous forests. *Forest Science* 47: 419–427.

Boring, L. R. 1979. Early forest regeneration and nutrient conservation on a clearcut Southern Appalachian watershed. Thesis. University of Georgia, Athens.

Boring, L. R. 1982. The role of black locust (*Robinia pseudoacacia* L.) in forest regeneration and nitrogen fixation in the southern Appalachians. Dissertation. University of Georgia, Athens.

Boring, L. R., C. D. Monk, and W. T. Swank. 1981. Early regeneration of a clear-cut Southern Appalachian forest. *Ecology* 62: 1244–1253.

Boring, L. R., and W. T. Swank. 1984. The role of black locust (*Robinia pseudoacacia*) in forest succession. *Journal of Ecology* 72: 749–766.

Boring, L. R., and W. T. Swank. 1986. Hardwood biomass and net primary production following clearcutting in the Coweeta Basin. Pages 43–50 in *Proceedings of the 1986 Southern Forest Biomass Workshop.* R. T. Brooks, Jr., editor. Tennessee Valley Authority, Norris, Tennessee.

Boring, L. R., D. L. White, and B. L. Haines. 1987. Litterfall and throughfall nitrogen transfers in black locust and pine-hardwood stands. *Nitrogen Fixing Tree Research Reports* 5: 54–56.

Boring, L. R., W.T. Swank, and C. D. Monk. 1988. Dynamics of early successional forest structure and processes in the Coweeta Basin. Pages 161–180 in *Forest Hydrology and Ecology at Coweeta.* W. T. Swank and D. A. Crossley, Jr., editors. Springer-Verlag, New York, New York.

Bormann, F. H., G. E. Likens, T. G. Siccama, R. S. Pierce, and J. S. Eaton. 1974. The export of nutrients and recovery of stable conditions following deforestation at Hubbard Brook. *Ecological Monographs* 44: 255–277.

Bormann, F. H., G. E. Likens, and J. M. Melillo. 1977. Nitrogen budget for an aggrading northern hardwood forest ecosystem. *Science* 196: 981–983.

Bormann, F. H., and G. E. Likens. 1979. *Pattern and Process in a Forested Ecosystem.* Springer-Verlag, New York, New York.

Britton, K. O. 1993. Anthracnose infection of dogwood seedlings exposed to natural inoculum in western North Carolina. *Plant Disease* 77: 34–37.

Brosofske, K. D., J. Chen, and T. R. Crow. 2001. Understory vegetation and site factors: implications for a managed Wisconsin landscape. *Forest Ecology and Management* 146: 75–87.

Brown, D. 1974. The development of woody vegetation in the first 6 years following clearcutting of a hardwood forest for a utility right-of-way. *Forest Ecology and Management* 65: 171–181.

Brown, S. L., P. Schroeder, and J. S. Kern. 1999. Spatial distribution of biomass in forests of the eastern USA. *Forest Ecology and Management* 123: 81–90.

Buckner, E., and W. McCracken. 1978. Yellow poplar: a component of climax forests. *Journal of Forestry* 76: 421–423.

Burns, R.M., and B. H. Honkala. 1990. Silvics of North America. 2. Hardwoods. Agriculture Handbook 654. USDA Forest Service, Washington DC.

Busing, R. T. 1989. A half century of change in a Great Smoky Mountains cove forest. *Bulletin of the Torrey Botanical Club* 116: 283–288.

Busing, R. T., E. E. C. Clebsch, and P. S. White. 1993. Biomass and production of southern Appalachian cove forests reexamined. *Canadian Journal of Forest Research* 23: 760–765.

Cannell, M. G. R. 1982. *World Forest Biomass and Primary Production Data*. Academic Press, New York, New York.

Canham, C. D., and Marks, P. L. 1985. The response of woody plants to disturbance: patterns of establishment and growth. Pages 197–216 in *The Ecology of Natural Disturbance and Patch Dynamics*. S. T. A. Pickett and P. S. White, editors. Academic Press, New York, New York.

Chellemi, D. O., K. O. Britton, and W. T. Swank. 1992. Influences of site factors on dogwood anthracnose in the Nantahala Mountain Range of western North Carolina. *Plant Disease* 76: 915–918.

Clinton, B. D., L. R. Boring, and W. T. Swank. 1993. Characteristics of drought-induced canopy gaps in oak forests of the Coweeta Basin. *Journal of Ecology* 74: 1551–1558.

Cole, D. W., and Rapp, M. 1981. Elemental cycling in forest ecosystems. Pages 341–409 in *Dynamic Principles of Forest Ecosystems*. D. E. Reichle, editor. Cambridge University Press, New York, New York.

Collins, B. S., K. P. Dunne, and S. T. A. Pickett. 1985. Response of forest herbs to canopy gaps. Pages 217–234 in *The Ecology of Natural Disturbance and Patch Dynamics*. S. T. A. Pickett and P. S. White, editors. Academic Press, New York, New York.

Cromack, K., Jr. 1973. Litter production and decomposition in a mixed hardwood watershed and a white pine watershed at Coweeta Hydrologic Station, North Carolina. Dissertation. University of Georgia, Athens.

Crow, T. R. 1978. Biomass and production in three contiguous forests in northern Wisconsin. *Ecology* 59: 265–273.

Crow, T. R., G. D. Mroz, and M. R. Gale. 1991. Regrowth and nutrient accumulations following whole-tree harvesting of a maple-oak forest. *Canadian Journal of Forest Research* 21: 1305–1315.

Crowell, M. and B. Freedman. 1994. Vegetation development in a hardwood-forest chronosequence in Nova Scotia. *Canadian Journal of Forest Research* 24: 260–271.

Day, F. P., and C. D. Monk. 1974. Vegetation patterns on a Southern Appalachian watershed. *Journal of Ecology* 55: 1064–1074.

Day, F. P., D. L. Phillips, and C. D. Monk. 1988. Forest communities and patterns. Pages 141–150 in *Forest Hydrology and Ecology at Coweeta*. W. T. Swank and D. A. Crossley, Jr., editors. Springer-Verlag, New York, New York.

DeLucia, E. H., T. W. Sipe, J. Herrick, and H. Maherali. 1998. Sapling biomass allocation and growth in the understory of a deciduous hardwood forest. *American Journal of Botany* 85: 955–963.

Dobbs, M. M. 1998. Dynamics of the evergreen understory at Coweeta Hydrologic Laboratory, North Carolina. Dissertation. University of Georgia, Athens.

Dobbs, M. M., and A. J. Parker. 2004. Evergreen understory dynamics in Coweeta forest, North Carolina. *Physical Geography* 25: 481–498.

Douglass, J. E., and Hoover, M. D. 1988. History of Coweeta. Pages 17–34 in *Forest Hydrology and Ecology at Coweeta*. W. T. Swank and D. A. Crossley, Jr., editors. Springer-Verlag, New York, New York.

Elliott, K. J., and W. T. Swank. 1994a. Changes in tree species diversity after successive clearcuts in the Southern Appalachians. *Vegetatio* 115: 11–18.

Elliott, K. J., and W. T. Swank. 1994b. Impacts of drought on tree mortality and growth in a mixed hardwood forest. *Journal of Vegetation Science* 5: 229–236.

Elliott, K. J., L. R. Boring, W. T. Swank, and B. R. Haines. 1997. Successional changes in plant species diversity and composition after clearcutting a southern Appalachian watershed. *Forest Ecology and Management* 92: 67–85.

Elliott, K. J., L. R. Boring, and W. T. Swank. 1998. Changes in vegetation structure and diversity after grass-to-forest succession in a southern Appalachian watershed. *American Midland Naturalist* 140: 219–232.

Elliott, K. J., J. M. Vose, W. T. Swank, and P. V. Bolstad. 1999. Long-term patterns in vegetation-site relationships in southern Appalachian forests. *Journal of the Torrey Botanical Society* 126: 320–334.

Elliott, K. J., L. R. Boring, and W. T. Swank. 2002. Aboveground biomass and nutrient pools in a Southern Appalachian watershed 20 years after clearcutting. *Canadian Journal of Forest Research* 32: 667–683.

Elliott, K. J., and W. T. Swank. 2008. Long-term changes in forest composition and diversity following early logging (1919-1923) and the decline of American chestnut (*Castanea dentata*). *Plant Ecology* 197: 155–172.

Elliott, K. J., and J. M. Vose. 2012. Age and distribution of an evergreen clonal shrub in the Coweeta Basin: *Rhododendron maximum* L. *Journal of the Torrey Botanical Society* 139: 149–166.

Gilliam, F. S., N. L. Turrill, and M. B. Adams. 1995. Herbaceous-layer and overstory species in clear-cut and mature central Appalachian hardwood forests. *Ecological Applications* 5: 947–955.

Gilliam, F. S. 2002. Effects of harvesting on herbaceous layer diversity of a central Appalachian hardwood forest in West Virginia, USA. *Forest Ecology and Management* 155: 33–43.

Gholz H. L., G. M. Hawk, A. Campbell, and K. Cromack. 1985. Early vegetation recovery and element cycles on a clear-cut watershed in western Oregon. *Canadian Journal of Forest Research* 18: 1427–1436.

Goodale, C. L., and J. D. Aber 2001. The long-term effects of land-use history on nitrogen cycling in northern hardwood forests. *Ecological Applications* 11: 253–267.

Gove, J. H., C. W. Martin, G. P. Patil, D. S. Solomon, and J. W. Hornbeck. 1992. Plant species diversity on even-aged harvests at the Hubbard Brook Experimental Forest: 10-year results. *Canadian Journal of Forest Research* 22: 1800–1806.

Grier, C. C., K. M. Lee, N. M. Nadkarni, G. O. Klock, and P. J. Edgerton. 1989. Productivity of forests of the United States and its relation to soil and site factors and management: a review. USDA Forest Service General Technical Report PNW-GTR-222.

Grimm, A. C. 1988. Fine root decomposition and nitrogen dynamics in early successional southern Appalachian forest. Thesis. Emory University, Atlanta, Georgia.

Hedman, C. W., and D. H. Van Lear. 1995. Vegetative structure and composition of southern Appalachian riparian forests. *Bulletin of the Torrey Botanical Club* 122: 134–144.

Helvey, J. D., and J. H. Patric. 1988. Research on interception losses and soil moisture relationships. Pages 126–138 in *Forest Hydrology and Ecology at Coweeta*. W. T. Swank and D. A. Crossley, Jr., editors. Springer-Verlag, New York, New York.

Hewlett, J. D. 1979. Forest water quality: an experiment in harvesting and regenerating Piedmont forests. Georgia Forest Research Paper.

Hornbeck, J. W., C. W. Martin, R. S. Pierce, F. H. Bormann, G. E. Likens, and J. S. Eaton. 1987. The northern hardwood forest ecosystem: ten years of recovery from clear-cutting. USDA Forest Service Research Paper NE-RP-596.

Hursh, C. R., and F. W. Haasis. 1931. Effects of 1925 summer drought on southern Appalachian hardwoods. *Ecology* 12: 380–386.

Jenkins, J. C., R. A. Birdsey, and Y. Pan. 2001. Biomass and NPP estimation for the mid-Atlantic region (USA) using plot-level forest inventory data. *Ecological Applications* 11: 1174–1193.

Johnson, E. A. 1952. Effect of farm woodland grazing on watershed values in the Southeast. *Journal of Forestry* 50: 109–113.

Johnson, P. L., and W. T. Swank. 1973. Studies of cation budgets in the Appalachians of four experimental watersheds with contrasting vegetation. *Ecology* 54: 70–80.

Kays, J. S., and C. D. Canham. 1992. Effects of time and frequency of cutting on hardwood root reserves and sprout growth. *Forest Science* 37: 524–539.

Knoepp, J. D., D. C. Coleman, D. A. Crossley, and J. S. Clark. 2000. Biological indices of soil quality: an ecosystem case study of their use. *Forest Ecology and Management* 138: 357–368.

Leopold, D. J., and G. R. Parker. 1985. Vegetation patterns on a southern Appalachian watershed after successive clearcuts. *Castanea* 50: 164–186.

Leopold, D. J., G. R. Parker, and W. T. Swank. 1985. Forest development after successive clearcuts in the southern Appalachians. *Forest Ecology and Management* 13: 83–120.

Likens, G. E., F. H. Bormann, R. S. Pierce, J. S. Eaton, and N. M. Johnson. 1977. *Biogeochemistry of a Forested Ecosystem*. Springer-Verlag, New York, New York.

Lipscomb, M. V., and E. T. Nilsen. 1990. Environmental and physiological factors influencing the natural distribution of evergreen and deciduous ericaceous shrubs on northeast and southwest slopes of the S. Appalachian Mountains. I. Irradiance tolerance. *American Journal of Botany* 77: 108–115.

Mann, L. K., D. W. Johnson, D. C. West, D. W. Cole, J. W. Hornbeck, C. W. Martin, H. Riekerk, C. T. Smith, W. T. Swank, L. M. Tritton, and D. H. Van Lear. 1988. Effects of whole-tree and stem-only clearcutting on postharvest hydrologic losses, nutrient capital, and regrowth. *Forest Science* 34: 412–428.

Martin, C. W., and A. S. Bailey. 1999. Twenty years of change in a northern hardwood forest. *Forest Ecology and Management* 123: 253–260.

Martin, C. W., and J. W. Hornbeck. 1989. Revegetation after strip cutting and block clearcutting in northern hardwoods: a ten-year history. USDA Forest Service Research Paper NE-RP-625.

Monk, C. D., and F. P. Day, Jr. 1988. Biomass, primary production, and selected nutrient budgets for an undisturbed watershed. Pages 151–159 in *Forest Hydrology and Ecology at Coweeta*. W. T. Swank and D. A. Crossley, Jr., editors. Springer-Verlag, New York, New York.

Monk, C. D., D. T. McGinty, and F. P. Day, Jr. 1985. The ecological importance of *Kalmia latifolia* and *Rhododendron maximum* in the deciduous forest of the Southern Appalachians. *Bulletin of the Torrey Botanical Club* 112: 193–197.

Monk, C. D., and F. P. Day, Jr. 1984. Vegetation analysis, primary production and selected nutrient budgets for a southern Appalachian oak forest: A synthesis of IBP studies at Coweeta. *Forest Ecology and Management* 10: 87–113.

Monk, C. D., G. I. Child, and S. A. Nicholson. 1970. Biomass, litter and leaf surface area estimates of an oak-hickory forest. *Oikos* 21: 138–141.

Montagnini, F., B. Haines, L. Boring, and W. Swank. 1986. Nitrification potentials in early successional black locust and mixed hardwood forest stands in the southern Appalachians. *Biogeochemistry* 2: 197–210.

Montagnini, F., B. L. Haines, W. T. Swank, and J. B. Waide. 1989. Nitrification in undisturbed mixed hardwoods and manipulated forests in the southern Appalachian Mountains of North Carolina. *Canadian Journal of Forest Research* 19: 1226–1234.

Nixon, E. S. and A. R. Brooks. 1991. Species diversity following clearcutting in eastern Texas. *Texas Journal of Science* 43: 399–403.

O'Neill, R. V., and D. L. DeAngelis. 1981. Comparative productivity and biomass relations of forest ecosystems. Pages 489–506 *in Dynamic Properties of Forest Ecosystems*. D. E. Reichle, editor. Cambridge University Press, London.

Parker, G. R., and W. T. Swank. 1982. Tree species response to clear-cutting a Southern Appalachian watershed. *American Midland Naturalist* 108: 304–310.

Phillips, D. L., and W. H. Murdy. 1985. Effects of rhododendron (*Rhododendron maximum* L.) on regeneration of southern Appalachian hardwoods. *Forest Science* 31: 226–233.

Phillips, D. L., and Shure, D. J. 1990. Patch-size effects on early succession in southern Appalachian forests. *Journal of Ecology* 71: 204–212.

Reich, P. B., D. F. Grigal, J. D. Aber, and S. T. Gower. 1997. Nitrogen mineralization and productivity in 50 hardwood and conifer stands on diverse soils. *Journal of Ecology* 78: 335–347.

Reiners, W. A. 1992. Twenty years of ecosystem reorganization following experimental deforestation and regrowth suppression. *Ecological Monographs* 62: 503–523.

Ross, M. S., T. L. Sharik, and D. W. Smith. 1986. Oak regeneration after clear felling in southwest Virginia. *Forest Science* 32: 157–169.

Seastedt, T. R., and D. A. Crossley, Jr. 1981. Microarthropod response following cable logging and clearcutting in the southern Appalachians. *Ecology* 62: 126–135.

Seastedt, T. R., D. A. Crossley Jr., V. Meentemeyer, and J. B. Waide. 1983a. A two-year study of leaf litter decomposition as related to macroclimatic factors and microarthropod abundance in the Southern Appalachians. *Holarctic Ecology* 6: 11–16.

Seastedt, T. R., D. A. Crossley, Jr., and W. W. Hargrove. 1983b. The effects of low-level consumption by canopy arthropods on the growth and nutrient dynamics of black locust and red maple trees in the southern Appalachians. *Ecology* 64: 1040–1048.

Shure, D. J., D. L. Phillips, and P. E. Bostick. 2006. Gap size and succession in cutover southern Appalachian forests: an 18-year study of vegetation dynamics. *Plant Ecology* 185: 299–318.

Smith, R. N. 1991. Species composition, stand structure, and woody detrital dynamics associated with pine mortality in the southern Appalachians. Thesis. University of Georgia, Athens.

Starkey, D. A., S. W. Oak, G. W. Ryan, F. H. Tainter, C. Redmond, and H. D. Brown. 1989. Evaluation of oak decline area in the South. USDA Forest Service Forest Protection Report R8-TR 17, Washington, DC.

Stringer, J. W., T. W. Kimmerer, J. C. Overstreet, and J. P. Dunn. 1989. Oak mortality in eastern Kentucky. *Southern Journal of Applied Forestry* 13: 86–91.

Swank, W. T., J. M. Vose, and K. J. Elliott. 2001. Long-term hydrologic and water quality responses following commercial clearcutting of mixed hardwoods on a southern Appalachian catchment. *Forest Ecology and Management* 133: 1–16.

Swank, W. T., and J. M. Vose. 1988. Effects of cutting practices on microenvironment in relation to hardwood regeneration. Pages 71–88 in *Guidelines for regenerating Appalachian Hardwood Stands: Proceedings of a Workshop*, Morgantown, West Virginia, 24-26 May 1988. H. C. Clay, A. W. Perkey, W. E. Kidd, Jr., editors. Society of American Foresters Publication 88-03, West Virginia University Books, Morgantown.

Swank, W. T., and W. H. Caskey. 1982. Nitrate depletion in a second-order mountain stream. *Journal of Environmental Quality* 11: 581–584.

Swank, W. T., and J. D. Helvey. 1970. Reduction of streamflow increases following regrowth of clearcut hardwood forests. Pages 346–360 in Symposium on the results of research on representative and experimental basins: 1970 December, Wellington, New Zealand, Publication 96. Leuven, Belgium: United Nations Educational, Scientific and Cultural Organization—International Association of Scientific Hydrology.

Swift, L. W., Jr., G. B. Cunningham, and J. E. Douglass. 1988. Climatology and hydrology. Pages 35–55 in *Forest Hydrology and Ecology at Coweeta*. W. T. Swank and D. A. Crossley, Jr., editors. Springer-Verlag, New York, New York.

Swift, L. W., Jr., and W. T. Swank. 1981. Long term responses of streamflow following clearcutting and regrowth. *Hydrological Sciences Bulletin* 26: 245–256.

Tritton, L. M., C. W. Martin, J. M. Hornbeck, and R. S. Pierce. 1987. Biomass and nutrient removals from commercial thinning and whole-tree clearcutting of central hardwood. *Envrionmental Management* 11: 659–666.

Tainter, F. H., S. W. Fraedrich, and D. M. Benson. 1984. The effect of climate on growth, decline and death of northern red oaks in the western North Carolina Nantahala Mountains. *Castanea* 49: 127–137.

Van Lear, D. H. 1991. Fire and oak regeneration in the southern Appalachians. Pages 15–21 in USDA Forest Service General Technical Report SE-69. Southeastern Forest Experiment Station, Asheville, North Carolina.

Van Lear, D. H., J. E. Douglass, S. K. Cox, and M. K. Augspurger. 1985. Sediment and nutrient export in runoff from burned and harvested pine watersheds in the South Carolina piedmont. *Journal of Environmental Quality* 14: 169–174.

Vose, J. M., and W. T. Swank. 1993. Site preparation burning to improve southern Appalachian pine-hardwood stands: aboveground biomass, forest floor mass, and nitrogen and carbon pools. *Canadian Journal of Forest Research* 23: 2255–2262.

Waide, J. B., W. H. Caskey, R. L. Todd, and L. R. Boring. 1988. Changes in soil nitrogen pools and transformations following forest clearcutting. Pages 221–232 in *Forest Hydrology and Ecology at Coweeta*. W. T. Swank and D. A. Crossley, Jr., editors. Springer-Verlag, New York, New York.

Wang, Z., and R. D. Nyland. 1993. Tree species richness increased by clearcutting of northern hardwoods in central New York. *Forest Ecology and Management* 57: 71–84.

Waring, R. H., and W. H. Schlesinger. 1985. *Forest Ecosystems Concepts and Management*. Academic Press, New York, New York.

Waring, R. H., J. J. Landsberg, and M. Williams. 1998. Net primary production of forests: a constant fraction of gross primary production? *Tree Physiology* 18: 129–134.

White, A. S. 1991. The importance of different forms of regeneration to secondary succession in a Maine hardwood forest. *Bulletin of the Torrey Botanical Club* 118: 303–311.

White, D. L. 1986. Litter production, decomposition, and nitrogen dynamics in black locust and pine-hardwood stands of the southern Appalachians. Thesis. University of Georgia, Athens.

White, D. L., B. L. Haines, and L. R. Boring. 1988. Litter decomposition in southern Appalachian black locust and pine-hardwood stands: litter quality and nitrogen dynamics. *Canadian Journal of Forest Research* 18: 54–63.

Williams, J. G. 1954. A study of the effect of grazing upon changes in vegetation on a watershed in the southern Appalachian mountains. Thesis. Michigan State College of Agriculture and Applied Science, East Lansing.

Woods, F. W., and R. E. Shanks. 1959. Natural replacement of chestnut by other species. *Ecology* 40: 349–361.

Yeakley, J. A., W. T. Swank, L. W. Swift, Jr., G. M. Hornberger, and H. H. Shugart. 1998. Soil moisture gradients and controls on a southern Appalachian hillslope from drought through recharge. *Hydrology and Earth System Science* 2: 31–39.

Yorks, T. E., and S. Dabydeen. 1999. Seasonal and successional understory vascular plant diversity in second-growth hardwood clearcuts of western Maryland. *Forest Ecology and Management* 119: 217–230.

3

Response and Recovery of Water Yield and Timing, Stream Sediment, Abiotic Parameters, and Stream Chemistry Following Logging

Wayne T. Swank*
Jennifer D. Knoepp
James M. Vose
Stephanie N. Laseter
Jackson R. Webster

Introduction

Watershed ecosystem analysis provides a scientific approach to quantifying and integrating resource responses to management (Hornbeck and Swank 1992) and also to address issues of resource sustainability (Christensen et al. 1996). The philosophical components of the research approach at Coweeta are (1) the quantity, timing, and quality of streamflow provides an integrated measure of ecosystem response to land management practices and (2) response to disturbance provides a valuable tool for interpreting ecosystem behavior (Swank and Crossley 1988).

Our objectives in this chapter are to (1) summarize and evaluate the long-term hydrologic and water quality responses to forest management and (2) link stream responses with process level research conducted within the watershed.

The details of general and specific forest study sites, experimental design, management prescriptions, and natural disturbances spanning the 32-year history of the study at Coweeta are described by Swank and Webster in chapter 1 of this volume. Briefly, a 59-ha south-facing mixed hardwood covered watershed was clearcut and logged in 1977 using a mobile cable system that could access logs up to 300 m from a road and suspend the logs completely above the ground for transport to the logging deck. Watershed (WS) 2, a 12.6-ha watershed adjacent to WS 7 served as the experimental control for assessing hydrologic and water quality responses to the treatment on WS 7.

* Corresponding author: Coweeta Hydrologic Laboratory, USDA Forest Service, 3160 Coweeta Lab Road, Otto, NC 28763 USA

Hydrology

Methods

Precipitation Input

Precipitation inputs were measured using standard rain gages located within the Coweeta basin. Precipitation input for each watershed was calculated using established relationships between specific watershed locations within the basin and individual or multiple rain gages.

Streamflow and Annual Water Yield

We used the paired or control catchment method of analysis (Hewlett et al. 1969) to quantify the effects of logging treatment on the quantity, timing, and quality of streamflow. In this method, the relationship of stream attributes between reference and treated watersheds for the calibration period is determined by regression analysis which incorporates experimental control for climatic and biological variations within and between years. The calibration period for hydrologic analysis in this study spanned 11 years, from 1966 to 1976, with continuous measurement of discharge using sharp-crested V-notch weirs (figure 3.1). Mean annual discharge from WS 7 during this period averaged 106 cm and ranged from 76 to 149 cm.

Figure 3.1   Upstream view of 90° V-notch weir installation on WS 7, November 2012. (USDA Forest Service photo)

The coefficient of determination (r^2) for total annual flow between WS 7 and WS 2 during the calibration period was 0.99. The error term ($p < 0.05$) for predicted individual annual flows for treatment years averaged ±5 cm. Regression analysis using monthly flow data was used to quantify the within-year changes; r^2 during the calibration period ranged from 0.96 to 0.99.

Early Postharvest Slash Interception Loss

An important component of evapotranspiration in forests is interception loss, which has seldom been studied after logging. Forest canopies intercept and alter the amount and chemistry of precipitation as water passes through foliage (throughfall) or flows down the stem (stemflow). Clearcutting and harvest on WS 7 removed the canopy structure for several years and added a large quantity of woody residue to the forest floor. Several (1 yr and 8 yr after cutting) studies were conducted to quantify the effects on interception loss (this chapter) and on nutrient leaching (see Knoepp et al., chapter 4, this volume).

The basic experimental design of the first study was the establishment of eighteen 4 x 4 m plots at 9 locations in WS 7, which were stratified to proportionally represent the forest types before clearcutting. One 2 x 2 m plot was nested in one corner of each 4 x 4 m plot; thus, there were a total of 36 plots. In the first year after cutting, coarse wood (CW; i.e., logs and branches with diameters ≥ 5 cm) was measured for end diameters and length of CW lying within each 4 x 4 m plot to calculate volume and surface area. Disks were taken from representative logs to determine wood density and mass. CW sampling was repeated in years 6, 7, and 11 in a long-term study of wood decomposition on WS 7 (Mattson et al. 1987; see also Mattson and Swank, chapter 7, this volume).

Fine wood (branches and stems < 5 cm in diameter) was sampled in the 2 x 2 m plots nested in the corner of each CW plot. These plots were selected to represent the range of slash dominated by stems, "brush," or mixed slash, all < 5 cm in diameter. Wood was also sampled on these plots to estimate surface area and biomass of fine wood.

Throughfall was collected by inserting 15 x 200 cm V-shaped aluminum troughs beneath the slash, attached to 19-L polypropylene collection jugs (see figure 4.1, in chapter 4, this volume). Samples were collected on a storm event basis and volumes used to estimate interception loss and leaching of nutrients from slash.

Later Postharvest Forest Interception Loss

A detailed study of interception loss, atmospheric deposition, and foliage leaching was conducted on WS 7 when the regenerating forest was 8 years old (Potter et al. 1991; Potter 1992) (figure 3.2). Interception findings are reported here and canopy nutrient fluxes are reported by Knoepp et al. (see chapter 4, this volume). Three 20 x 20 m plots were located near the middle of WS 7 in a chestnut oak (*Quercus prinus*) community. A 10-m tower was located in the middle of the study site and instrumented to collect incident rainfall and dry particulate inputs to the canopy.

Figure 3.2   WS 7 eight years after clearcutting. (USDA Forest Service photo)

Thirty troughs (1.0 x 0.1 m) were randomly placed in the three plots to collect throughfall; stemflow was measured in nine 1 x 2 m plots (Potter et al. 1991). Data were collected on a storm-event basis and included 20 storms, 14 during the growing season and 6 during the dormant season, throughout the period July 1984 through August 1986.

Results and Discussion

Initial Water Yield and Interception Responses

In 1978, the first full water year (May to April) following logging, streamflow increased 26.5 cm, or about 28% above the flow expected if the forest had not been cut (figure 3.3). In subsequent years, annual discharge increases declined at a rate of 5 to 7 cm per year until the fifth year after cutting, when annual flow was just 4 cm above pretreatment levels. Thereafter, changes in flow were not significant (p > 0.05) and discharge fluctuated around expected baseline values.

The pattern of initial response and early recovery of annual streamflow after clearcutting WS 7 are consistent with other forest cutting experiments at Coweeta (Swank et al. 1988) and in other locations of the Appalachian region of the United States (see Adams and Kochendenfer, chapter 12 and Hornbeck et al., chapter 13, this volume). Water yield increases are typically greatest in the first year after cutting because transpiration is most reduced due to minimal leaf area index (LAI). In subsequent years, as sprouts and seedlings regrow, LAI and transpiration increase (see Boring et al., chapter 2, this volume) resulting in a logarithmic decline in streamflow over the first six years of succession.

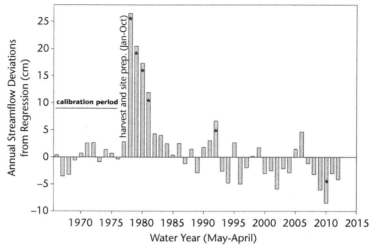

Figure 3.3 Annual deviations in streamflow on WS 7 prior to and following clearcutting and commercial logging at Coweeta Hydrologic Laboratory, 1967–2012. * Denotes significant change; p < 0.05).

Logging slash interception was measured for a total of 36 storms, ranging from 5 to 12 mm from December 1977 through April 1979. Interception loss (precipitation minus throughfall) was a linear function of rainfall amount for all types of slash. Statistical analyses of regression slopes showed no significant differences among the slash types < 5 cm, so all data from the small material were combined in the regressions. However, the regression slope of the coarse wood (logs and branches ≥ 5 cm) was different from all of the other types of slash and resulted in a different regression equation.

Linear regressions for estimating interception loss from slash are:

$$\text{IL (CWD)} = -0.44 + 0.2098 \text{ (P)}; \ r^2 = 0.70 \qquad \text{(Eq. 1)}$$
$$\text{IL (other slash)} = -0.82 + 0.1516 \text{ (P)}; \ r^2 = 0.47 \qquad \text{(Eq. 2)}$$

where IL is interception loss and P is precipitation, both in mm.

Estimates of annual interception loss using Eqs. 1 and 2 were derived for the period September 1977–August 1978; storm precipitation was measured monthly on a gage located on WS 7 and adjusted for each month based on previously developed seasonal weighting factors (Swift et al. 1988). Precipitation for the 12-month period totaled 1,731 mm, with a monthly range of 22 to 265 mm delivered in 76 storms over the year. Interception loss for CW was estimated to be 324 mm, or 18.7% of precipitation, compared to an estimated interception loss of 197 mm, or 11.4% of precipitation for smaller slash. A combined wood interception loss was derived by weighting the amount of the two slash types measured on the 36 plots based on wood surface area. The mean wood surface area per plot area was 0.448 m^2/m^2 for CW and 0.898 m^2/m^2 for smaller slash, for a total wood surface area of

1.346 m^2/m^2. Using the appropriate weighting factors, annual interception loss for wood was estimated to be 239 mm, or 13.8% of precipitation.

Similar studies of interception by logging residue are not available for comparison with these findings. However, Helvey and Patric (1988) reported an annual interception loss for mature hardwoods at Coweeta of 250 mm, or 13% total precipitation, including 3% litter interception loss. Thus, the early postharvesting interception component of the hydrologic budget on WS 7 was not substantially altered and the first several years of annual water yield increases measured after cutting and harvest (figure 3.3) were mainly due to reductions in transpiration. There are several factors contributing to the initial high interception loss. First, only sawlogs were harvested and the remaining vegetation was cut as part of the site preparation treatment, which contributed to a high loading of woody residue. Secondly, WS 7 is a south-facing slope; previous research (Swift 1972) found radiation available for evapotranspiration is greater on south- than north-facing slopes, an important factor that was included in a model used to predict water yield response on WS 7 (table 3.1).

Later Postharvest Forest Interception Loss

In the regenerating 8-year-old forest (Potter et al. 1991; Potter 1992), data were combined for all storms. Throughfall and stemflow were estimated to be 83.3% and 5.8% of precipitation, respectively, for an estimated interception loss of 10.9% of precipitation. These water fluxes for the regenerating stand were similar in magnitude to those that were concurrently measured in the adjacent reference hardwood forest on WS 2 (Swank et al. 1992).

The recovery of canopy interception loss early in succession can be mainly attributed to a rapid recovery of leaf area on WS 7 (see Boring et al., chapter 2, this volume). The contributions of interception loss to watershed evaporotranspiration (Et) and streamflow recovery is a dynamic process. The importance of interception loss from wood measured the first year after cutting declined over time with decomposition. However, 11 years after cutting, the quantity of CW was still significant (see Mattson and Swank, chapter 7, this volume), which contributed to

Table 3.1 Comparison of annual observed vs. predicted increase in water yield following clearcutting on Coweeta WS 7.

Year following clearcutting	Increased predicted by model (cm)	Observed Increase (cm)
1	25	26
2	17	20
3	12	17
4	8	12
5	6	4
6	3	4
Total	71	83

Source: Swank et al. (1988).

water storage and interception loss from the forest floor. Over the same time period, canopy interception loss was returning to precutting levels. Thus, during the first decade of regrowth, the combined interception loss from wood and the regrowing canopy probably exceeded that for a mature hardwood forest and may have contributed to some of the variability in annual water yield (figure 3.3).

Longer-Term Water Yield Responses

The significant increase in streamflow in 1992, when the forest was 15 years old (figure 3.3) has been attributed to a reduction in both stem density and LAI associated with competition and self-thinning (stem exclusion stage) of rapid growing coppice vegetation (Swank et al 2001; Elliott et al. 1997), a high mortality rate of *Robinia pseudoacacia* caused by stem borers (see Boring et al, chapter 2, this volume), decline of dogwood (*Cornus florida*) due to a disease, dogwood anthracnose (Chellemi et al. 1992), and loss of abundant American chestnut (*Castanea dentata*) sprouts due to the chestnut blight (see Boring et al., chapter 2, this volume). Similar patterns of changes in stand structure, water use, and streamflow have been found in other clearcutting experiments at Coweeta (Swift and Swank 1981).

Canopy openings created during the stem exclusion stage of succession were short-lived; and by 1994 to 1995, 17 years after cutting, the stand basal area was 23 m²/ha, which is similar to the 25 m²/ha basal area of the original forest (Elliott et al. 1997) and LAI also increased to precutting levels (see Boring et al., chapter 2, this volume). In the ensuing decade after LAI stabilized, there has been a pattern of annual streamflow reductions that frequently exceed 2.5 cm (figure 3.3). We hypothesize that higher Et for the regrowing versus mature forest is related to higher transpiration loss associated with major shifts in species composition. For example, there were very large increases in basal area of *Liriodendron tulipifera*, *Acer rubrum*, and *Robinia pseudoacacia* in the successional forest and an equally large decline for combined *Carya* and *Quercus* spp. (see Boring et al., chapter 2, this volume). This hypothesis is supported by recent physiological research at Coweeta that has shown large differences in canopy transpiration rates among hardwood species. Specifically, diffuse porous species, such as yellow popular and red maple, have much higher transpiration rates compared to oak and hickory species (Ford et al. 2010; Ford et al. 2011). Persistent decreases in annual water yield were also observed at Hubbard Brook following a harvest and attributed to higher transpiration rates for ring porous early succession species compared to mature northern hardwood forests (Hornbeck et al. 1977; see also Hornbeck et al., chapter 13, this volume).

Water Yield Model

One of the original objectives of the cutting experiment on WS 7 was to obtain water yield response data for a south-facing watershed. Previous syntheses of watershed experiments in the Appalachian Highlands Physiographic region (Douglass and Swank 1972, 1975) established empirical equations between first-year water yield increases as a function of percent basal area cut and an insolation index (energy variable related to slope aspect) for a catchment. In addition, another empirical equation was developed (Douglass and Swank 1975) for predicting water yield increases for any year following harvest until streamflow returns to baseline levels. The models contain little data from south-facing clearcut watersheds with natural forest succession. However, predictions from these models generally showed good agreement with annual water yield responses measured on south-facing WS 7 (table 3.1). The first year after cutting, streamflow increased about 26 cm compared to 25 cm predicted by the model. Subsequently, in the next 2 years, predicted increases were substantially below observed increases. These years coincided with the wettest year on record and one of the driest years at Coweeta. In the ensuing 3 years, predictions were in close agreement with observed increases in water yield and the total change predicted for the 6-year period was within 17% of the observed change.

Intra-Annual Water Yield

During the first 3 postcutting years, the monthly distributions of water yield increases (figure 3.4) were similar to other low-elevation cutting experiments at Coweeta (Swank et al. 1988). Flow increases occurred in every month, with the smallest amounts in the spring (April and May), at the same time that soil moisture in an undisturbed forest is usually fully recharged. Concurrent with the growing season, streamflow increases become larger due to reduced Et on WS 7. Substantial flow increases continued into the late fall and winter months and partially reflect the lag between when reduced Et occurs on the cut watershed and when the water savings reach the weir during the period of high precipitation (figure 3.4).

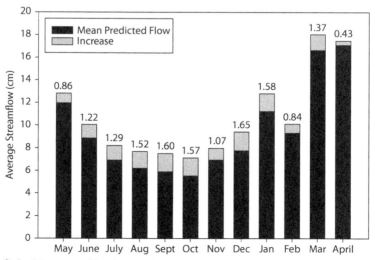

Figure 3.4   Mean monthly changes in water yield during the first 4 postcutting years on Coweeta WS 7. (Modified from Swank et al., 1982)

Of notable importance is that two of the largest flow increases (1.6 cm, or 28% increase) occurred in September and October when flows are normally low and water demands are high.

Storm Hydrograph Responses

Detailed analysis of eight storm hydrograph parameters was conducted for WS 7 using pretreatment storm data on WS 7 regressed against the same parameters as the reference catchment (WS 2). The analysis used data for 75 storms (≥ 2 cm) from the first 4 years after treatment, which encompassed the period of maximum water yield increase (Swank et al. 2001).

Following harvest, statistically significant changes in regression intercepts and slopes were found for all storm parameters except time to storm peak (table 3.2). The largest increases occurred in peak flow rates (15%) and initial flow rates (14%); the latter is due to elevated rates of baseflow from the watershed. Quickflow (stormflow) volume increased 10%, which was associated with a 10% increase in recession time. Taken collectively, the hydrograph responses are considered to be of minor importance to downstream flooding. For example, in the first 4 years after cutting and harvest, the average precipitation storm, ≥ 2 cm, increased the quickflow volume by only 0.03 cm, or 2.43 m³/ha, and the peakflow rate by 1.7 m³/ha.

Storm hydrograph responses to harvest are partly dependent upon (1) the magnitude and method of logging and associated road disturbance and (2) the inherent responsiveness of the watershed to precipitation events in the absence of

Table 3.2   Storm hydrograph parameters and changes during the first 4 years following clearcutting and logging on WS 7.

Parameter	Mean for treatment watershed		Significance of regression coefficients		Percent change in parameter after treatment for mean storm
	WS 7	WS 2	Intercept	Slope	
Initial flow rate (m³ s⁻¹ km⁻²)	0.037	0.27	**	**	14
Peakflow rate(m³ s⁻¹ km⁻²)	0.136	0.127	**	**	15
Time to peak (h)	8.0	8.0	NS[b]	NS	0
Total quickflow volume (cm)	0.32	0.40	**	**	10
Quickflow after peak (cm)	0.08	0.09	**	**	6
Quickflow after peak (cm)	0.24	0.31	*	NS	11
Quickflow duration (h)	27.6	28.3	**	*	5
Recession time (h)	20.0	21.0	**	NS	10

[a] Derived from difference between value predicted from calibration regression and measured value.
[b] Nonsignificant.
* p < 0.05.
** p < 0.01.
Source: Modified from Swank et al. (1982).

disturbance. Inherent responsiveness is driven by a variety of physical factors such as watershed size, soil depth, slope and topographic complexity, and infiltration rates. The response factor (mean annual quickflow/mean annual precipitation) for WS 7 was very low (0.04), which accounts for some of the small changes in storm hydrograph parameters. Furthermore, the low density of logging roads, minimal disturbance of the surface soil by cable logging (see Swank and Webster, chapter 1, this volume), and careful design of roads (Swift 1988) also limited changes in stormflow on WS 7.

Abiotic Responses to Clearcutting

Soil Moisture and Temperature

Regeneration cutting can produce significant changes in the microenvironment of the forest floor and soil that in turn regulate ecosystem processes, such as decomposition; microbial activity; nutrient cycles; and the germination, sprouting, survival, and growth of vegetation. Beginning in August 1977, studies were initiated on WS 7 to evaluate the effects of harvesting on soil moisture and temperature. Soil moisture was measured at biweekly intervals in the O_1 and O_2 (organic) litter layers of the forest floor and 0 to 10-cm and 10–30-cm depths in the soil on WS 7 and also WS 2, the adjacent control watershed (Swank and Vose 1988). In the first autumn (Aug–Nov) after harvest, litter moisture was 20% to 30% below that found on WS 2 (table 3.3). In the subsequent winter quarter, litter moisture was similar on both watersheds. However, in the ensuing year (1978), litter moisture was consistently 30% to 50 % lower in the clearcut compared to WS 2. In contrast, surface soil moisture (0–10 cm); increased more than 11% the first year after cutting (table 3.3). Deeper in the soil profile, moisture increases were small or showed no changes.

Soil temperatures were measured at the litter-soil interface on WS 7 during the precut year (1976) prior to clearcutting and in the first growing season after cutting and logging (Swank and Vose 1988). Mean monthly temperatures were 7°C to 10°C above precut levels in the period May through October in the first year after cutting. Mean monthly maximum soil temperatures showed large increases with values being 10°C to 35°C above precut levels and daily maximum temperatures frequently exceeded 45°C during this period. In subsequent years, increases in surface soil temperatures were moderated by shade from regeneration.

Stream Chemistry

Methods

Stream chemistry measurements began in late 1971 on both WS 7 and WS 2. Weekly grab samples have been collected at a fixed location just above the weir from each watershed since 1971, and flow proportional samples were also collected in the period 1975–1981. Solute determinations include NO_3^-, NH_4^+, SO_4^{-2}, PO_4^{-3}, Cl^-, and

Table 3.3   Quarterly forest floor and soil moisture averages (percent) for mixed hardwood forests during the first year after clearcutting (WS 7) and for the reference watershed (WS 2).

	Year and Quater			
	Aug–Nov 1977	Dec–Mar 1977–78	Apr–Jul 1978	Aug–Nov 1978
Treatment and depth	Water content (percent by weight)			
Clearcut				
O_1	70	124	88	51
O_2	95	170	97	58
0 to < 10 cm	46	70	55	31
10 to 30 cm	36	40	34	23
Control				
O_1	92	121	120	105
O_2	126	211	120	109
0 to < 10 cm	35	59	36	28
10 to 30 cm	32	40	29	24

Note: O_1 = includes fresh or slightly decomposed organic materials. O_2 = includes intermediate and highly decomposed organic materials.
Source: Modified from Swank and Vose (1988).

base cations (Ca^{+2}, Mg^{+2}, K^+, Na^+) using established analytical methods at Coweeta (Brown et al. 2009). Comparison of 3-year average annual export of solutes calculated from weekly grab samples versus flow-proportional samples showed good agreement for most solutes (Swank and Waide 1988). Annual change in export of each solute due to treatment the first 6 years was estimated from pretreatment regressions of monthly exports between WS 7 and WS 2. Relationships of monthly exports between the two catchments were good (r^2 values ≥ 0.92) for most ions.

Response to Treatment

Stream chemistry responses to treatment were relatively small (table 3.4) as described in an earlier analysis (Swank et al. 2001). Increases in export of PO_4, K, Ca, and Mg in the first full year following logging are partially related to release from the fertilizer applied to roads. Lack of significant NO_3 response was due in part to sediment denitrification that depleted NO_3 before it reached the weir (Swank and Caskey 1982). Denitrification in Big Hurricane Branch (WS 7) was remeasured in 2004 as part of a large regional study in 49 streams with varying land-use categories (Mulholland et al. 2009). They found measureable but lower rates of denitrification than those found in 1977 by Swank and Caskey (1982). However, differences in both methods and the supply of NO_3 of the two studies limit direct quantitative comparisons.

The magnitude of nutrient export is determined by both changes in solute concentrations and increases in discharge resulting from reduced Et following cutting. The first two years after cutting, annual flow increased an average of 23.5 cm/y, but

Table 3.4 Annual changes in streamflow and solutes following clearcutting and logging (posttreatment–pretreatment) on WS 7.

Time since treatment (May–April water year)	Flow (cm)	Increase or decrease in streamflow and solute export (kg ha^{-1})[a]								
		NO$_3$-N	NH$_4$-N	PO$_4$-P	K	Na	Ca	Mg	SO$_4$-S	Cl
First 4 months	0.5	0.01	< 0.01	0.01	0.43	0.42	0.24	0.26	0.39	0.68
First full year	26.5	0.26	0.03	0.04	1.98	1.37	2.60	0.96	0.27	1.13
Year 2	20.5	1.12	< 0.01	0.01	1.95	2.22	2.51	1.15	−0.08	1.62
Year 3	17.3	1.27	0.05	0.02	2.40	2.68	3.16	1.42	0.39	2.08
Year 4	11.9	0.25	0.15	0.02	0.80	1.07	1.63	0.46	0.31	0.59
Year 5	4.3	0.28	0.01	< 0.01	0.52	0.13	1.19	0.18	0.04	0.10
Year 6	4.1	0.62	0.06	< 0.01	0.73	0.69	0.89	0.42	−0.06	0.33

[a] Annual increase or decrease derived from sum of deviations using monthly calibration regressions.
Source: Modified from Swank et al. (2001).

maximum concentrations of most solutes did not occur until the third year when water yield on WS 7 was still more than 17 cm above pretreatment levels (table 3.1). By the sixth year of postdisturbance, streamflow was near pretreatment levels and solute exports also appeared to be approaching pretreatment levels. The longer-term responses of most solutes showed a similar pattern. For example, small (5 µeq/L) increases in Ca concentrations were observed after logging but later returned to expected pretreatment levels (figure 3.5). Similarly, following the initial increase in K concentrations, interannual concentrations of K after 1983 were highly variable, and there were no consistent differences in concentrations between WS 7 and WS 2 (figure 3.5). The same pattern was true for Mg (figure 3.6). There was little change in SO$_4$ concentrations on WS 7 after cutting, but beginning in 1989 there has been a consistent decline in SO$_4$ concentrations on both WS 7 and reference WS 2 (figure 3.6). Concentrations of NH$_4$ and PO$_4$ were low and almost identical for WS 7 and WS 2 during the entire period of record. In contrast, the long-term record for NO$_3$ showed interesting, significant dynamics following cutting and forest succession (figure 3.7).

The initial increase in NO$_3$ concentrations on WS 7 was attributed to increases in soil N pools and concentrations the first three years after logging (Waide et al. 1988; see also Knoepp et al., chapter 4, this volume). Decline in stream NO$_3$ concentrations until about 1987 are associated with the rapid sequestration and storage of nutrients in successional vegetation (Boring et al. 1988; see also Boring et al., chapter 2, this volume). However, major shifts in internal ecosystem N cycling are evident in a large, sustained pulse of NO$_3$ to the stream from about 1987 through 1997 (figure 3.7). Mean annual peak NO$_3$ concentrations 20 years after disturbance are about double the values in the early postharvest years. Thereafter, NO$_3$ concentrations declined about 8 µeq/L over a 5-year period, followed by another increase during the ensuing 5 years that reached a maximum of 12 µeq/L in 2008 (figure 3.7).

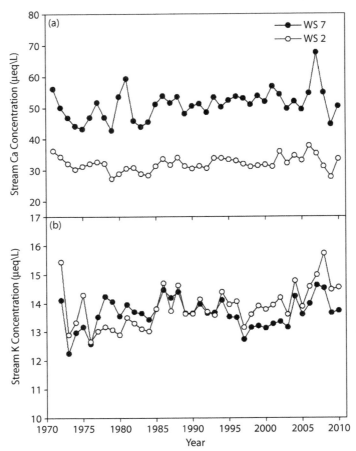

Figure 3.5 Mean annual concentrations of Ca (a) and K (b) in streamwater of WS 7 and WS 2 during calibration (1971–1976), treatment activities (1976–1977), and postharvest period (1978–2010).

A complex combination of ecological processes contribute to the magnitude and temporal dynamics in stream NO_3. Accelerated NO_3 release to the stream coincides with stem exclusion in 1992, and thus, some of the NO_3 loss was probably due to the reduced uptake. However, the largest contributor to increased N availability was extensive black locust mortality due to locust borer infestation, which is a similar response observed in another early succession watershed (WS 6) at Coweeta with large locust infestations and mortality (Swank and Waide 1988). Black locust is a symbiotic nitrogen fixer; in the 4-year-old locust stands on WS 7, fixation was estimated to be 30 kg N ha^{-1} yr^{-1} while fixation catchment wide was estimated at 10 kg N ha^{-1} yr^{-1} (Boring and Swank 1984). Moreover, black locust is known to accumulate large quantities of N in foliage, roots, branches, and stems (Boring and Swank 1984). Decomposition of the N-rich organic matter from the dead trees (330 stems/ha) could be a major source of stream NO_3 later in succession (figure 3.7).

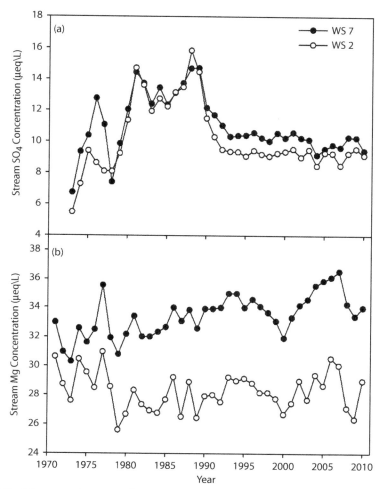

Figure 3.6   Mean annual concentrations of SO$_4$ (a) and Mg (b) in streamwater of WS 7 and WS 2 during calibration (1971–1976), treatment activities (1976–1977), and postharvest period (1978–2010).

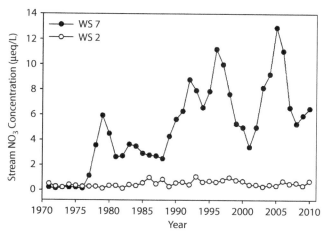

Figure 3.7   Mean annual concentrations of NO$_3$ in streamwater of WS 7 and WS 2 during calibration (1971–1978), treatment activities (1976–1977), and postharvest period (1978–2010).

Decomposing logging residue (see Mattson and Swank, chapter 7, this volume) is also a potential source of long-term stream NO_3 enrichment that appears in the stream. Other possible reasons for increased stream NO_3 concentrations are elevated rates of soil N mineralization and nitrification and reduction in the soil C/N ratio.

Measurement of the C/N ratio showed no significant change over the 18-year period following cutting (Knoepp and Swank 1997). However, long-term assessment of soil N transformations show continued increase in N availability in surface soils 20 years following harvest (see Knoepp et al., chapter 4, this volume).

The stream NO_3 responses observed on WS 7 demonstrate the value of long-term studies in forest ecosystems and the processes regulating system responses. For example, the first 10-year trend in concentrations suggested that NO_3 concentrations had returned to near pretreatment levels. However, in the subsequent 30-year period, concentrations showed very large increases and decreases that greatly exceeded the initial responses. Thus, early termination of the study and associated conclusions would have been incomplete and partly inaccurate.

The importance of the interaction between successional vegetation (e.g., black locust) and insect infestations on stream NO_3 is clearly evident on WS 7; however, this does not explain all of the temporal variation in stream NO_3. For example, interannual magnitude and variability of stream NO_3 concentration are also related to hydrologic variables. Annual streamflow on WS 7, ranged from 45 to 130 cm/y from 1990 to 2009, and explained 36% of the annual variation in stream NO_3 concentrations following cutting (figure 3.8).

Taken collectively, nutrient losses on WS 7 should not have an adverse impact on the sustainability and growth of the successional forest. Atmospheric deposition of nutrients exceeded the elevated losses of nutrients in most years of the study (Swank et al. 2001). Moreover, the high N-fixation rates of black locust and availability of N to other tree species can be viewed as a benefit to tree growth and forest health. Further discussion on the relevance of findings to management and ecological values is found in chapter 14 of this volume.

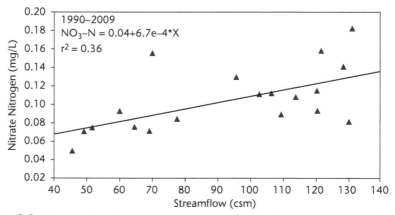

Figure 3.8   The relationship between mean annual NO_3-N concentrations and annual streamflow on Coweeta WS 7 over a 20-year period.

Sediment

Methods

The effects of the management practices on soil loss to streams on WS 7 were determined at periodic intervals by measuring sediment accumulations in the weir ponding basin and also on the approach apron of the ponding basin of both WS 7 and WS 2. This approach does not account for suspended sediment that passes across the weir blade; therefore, total sediment export was underestimated. Sediment volumes were estimated by measuring sediment elevations along permanent transects with a transit and level rod before and after cleaning the ponding basin and approach aprons. Bulk samples were collected at each elevation measurement and processed to estimate dry weight. A pretreatment calibration regression equation of periodic sediment loss over a 2-year period between WS 7 and WS 2 was derived ($r^2 = 0.91$) to estimate changes in sediment loss due to management.

Soil losses on subdrainages within WS 7 were also measured to separate and quantify sediment sources due to roads versus logging. Discharge and soil export were sampled with an H flume and a Coshocton wheel using procedures described by Douglass and Van Lear (1983). One installation was located in a perennial stream below the middle logging road (figure 3.9) and three installations were located above the influence of logging roads to evaluate effects of cutting and logging only.

Figure 3.9 One foot H-Flume and Coshocton sampler (no. 701) on WS 7—one of four such installations used to assess water quality above and below roads, 1976. (USDA Forest Service photo)

Results

Sediment yield in the 2 years of pretreatment calibration from WS 7 and WS 2 averaged 230 and 135 kg ha^{-1} yr^{-1} respectively. These baseline sediment yields are similar to the mean values for small, forested catchments in the eastern United States summarized by Patric et al. (1984). In mid-May 1976, roads in WS 7 were fertilized and seeded but road fills and the running surface were unsettled and without grass or gravel cover. The third week of May 1976, a 16-cm storm occurred and was followed May 28 by a larger storm of 22 cm with intensities of 7 cm/h. The second storm produced the greatest discharge rates measured on most catchments at Coweeta during the previous 62 years of the Laboratory gaging history. These two events greatly accelerated sediment yield on both WS 7 and WS 2 with an increase in soil loss in May on WS 7 of 1,470 kg ha^{-1} yr^{-1} (figure 3.10). Roads were the primary source of increased sediment yield as illustrated by soil loss measured at one of the gaging stations in a stream immediately below a road crossing the middle of the catchment (figure 3.11). Following the May storms, sediment yield at the station was nearly 50 t from 0.086 ha of road contributing area comprised of roadbed, cut, and fill. Following road stabilization and minimum use over the period of June through December 1976, soil loss from the road was low, but it accelerated again during the peak of logging activities (figure 3.10). In ensuing years, soil loss returned to baseline levels. In the same time period, samplers located above roads, which were only influenced by cutting and yarding, only collected small amounts of material comprised mainly of organic matter.

Following the initial pulse of sediment from the May 1976 storms, sediment yield showed much different temporal patterns at the weir (figure 3.10) compared to sediment loss from the roads. Sediment yield remained substantially elevated

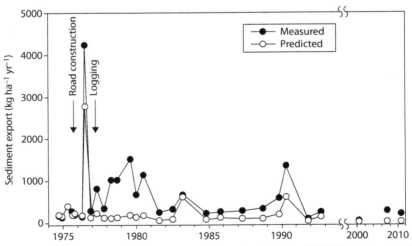

Figure 3.10 Sediment yield measured in the ponding basin on WS 7 following logging compared to the sediment yield predicted from WS 2, the adjacent reference watershed, over a 35-year period.

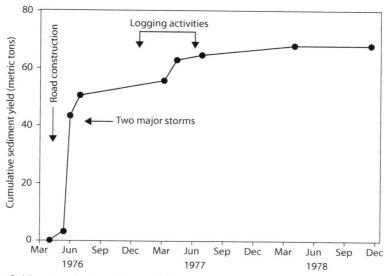

Figure 3.11   Cumulative sediment yield measured in a first-order stream below a logging road during the first 32 months after logging.

during and after logging disturbances and continued to have sediment losses that were frequently in excess of 500 kg ha^{-1} yr^{-1} above expected losses over the period 1978—1985 (figure 3.10). A large pulse of sediment was delivered to weirs on both WS 2 and WS 7 in 1989 in response to the wettest year on record at Coweeta. Sediment yield on WS 7 was about 800 kg ha^{-1} yr^{-1} above the expected yield. In 1991 sediment yield on WS 7 returned to pretreatment levels (234 kg ha^{-1} yr^{-1}) and remained at about the same level (179 kg ha^{-1} yr^{-1}) based on three sample periods in subsequent years (figure 3.10).

The long-term sediment yield responses illustrate the delay or lag between pulsed sediment inputs to a stream and routing of sediments through the water-shed. Soil loss derived from roads was very low following stabilization with grass and gravel cover. Moreover, following logging, road travel was minimal—roads were only used for access to research sites. Also, based on long-term cross-section measurements on the main stream on WS 7, there have only been infrequent and minor instances of stream bank erosion (Webster et al. unpublished data). Thus, in the absence of significant additional sources of sediment to WS 7 streams, the long-term sediment increases observed at the weir on WS 7 were due to a continued release of sediment from upstream storage, which was deposited from three road crossings on perennial streams and four crossings of intermittent streams on WS 7 in the May 1976 storms.

The unique conditions that produced these sediment responses should be recognized, that is, record storms occurred at the precise time when roads were freshly constructed and without vegetation cover, and thus most vulnerable to erosion. It is also important to point out that best management practices were used in harvesting and in logging road location and design. The long-term effect of this management prescription on water, soil, vegetation sustainability and health, and the structure and function of benthic invertebrates is further discussed in a concluding synthesis (see Webster et al., chapter 14, this volume).

Literature Cited

Boring, L. R., and W. T. Swank. 1984. The role of black locust (*Robina pseudo-acacia*) in forest succession. *Journal of Ecology* 72: 749–766.

Boring, L. R., W. T. Swank, and C. D. Monk. 1988. Dynamics of early successional forest structure and processes in the Coweeta basin. Pages 161–179 in *Forest Hydrology and Ecology at Coweeta*. W. T. Swank and D. A. Crossley, Jr., editors. Springer-Verlag, New York, New York.

Brown, C. L., C. Harper, N. Muldoon, and S. Cladis. 2009. Procedures for chemical analysis at the Coweeta Hydrologic Laboratory. Coweeta Files. Coweeta Hydrologic Laboratory, Otto, North Carolina.

Chellemi, D. O., K. O. Britton, and W. T. Swank. 1992. Influence of site factors on dogwood anthracnose in the Nantahala Mountain Range of Western North Carolina. *Plant Disease* 76: 915–918.

Christensen, N. L., A. M. Bartuska, J. H. Brown, S. Carpenter, C. D'Antonio, R. Francis, J. F. Franklin, J. A. MacMahon, R. F. Noss, D. J. Parsons, C. H. Peterson, M. G. Turner, and R. G. Woodmansee. 1996. The report of the Ecological Society of America Committee on the Scientific Basis for Ecosystem Management. *Ecological Applications* 6: 665–691.

Douglass, J. E., and W. T. Swank. 1972. Streamflow modification through management of eastern forests. Research Paper SE-94. USDA Forest Service, Southeastern Forest Experiment Station. Asheville, North Carolina.

Douglass, J. E., and W. T. Swank. 1975. Effects of management practices on water quality and quantity: Coweeta Hydrologic Laboratory, North Carolina. Pages 1–13 *in* Proceedings of the municipal watershed management symposium. General Technical Report NE-13. U.S. Department of Agriculture, Forest Service, Northeastern Forest Experiment Station, Broomall, Pennsylvania.

Douglass, J. E., and D. H. Van Lear. 1983. Prescribed burning and water quality at ephemeral streams in the Piedmont of South Carolina. *Forest Science* 29: 181–189.

Elliott, K. J., L. R. Boring, W. T. Swank, and B. R. Haines. 1997. Successional changes in plant species diversity and composition after clear-cutting a southern Appalachian watershed. *Forest Ecology and Management* 92: 67–85.

Ford, C. R., R. M. Hubbard, and J. M. Vose. 2010. Quantifying structural and physiological controls on variation in canopy transpiration among planted pine and hardwood species in the southern Appalachians. *Ecohydrology* 3. doi: 10.1002/eco.136

Ford, C. R., S. H. Laseter, W. T. Swank, and J. M. Vose. 2011. Can forest management be used to sustain water-based ecosystem service in the face of climate change? *Ecological Applications* 21: 2049–2067.

Helvey, J. D., and J. H. Patric. 1988. Research on interception losses and soil moisture relationships. Pages 129–137 in *Forest Hydrology and Ecology at Coweeta*. W. T. Swank and D. A. Crossley, Jr., editors. Springer-Verlag, New York, New York.

Hewlett, J. D., H. W. Lull, and K. G. Reinhart. 1969. In defense of experimental watersheds. *Water Resources Research* 5: 306–316.

Hornbeck, J. W., and W. T. Swank. 1992. Watershed ecosystem analysis as a basis for multiple-use management of eastern forests. *Ecological Applications* 2: 238–247.

Hornbeck, J. W., C. W. Martin, and C. Eagar. 1997. Summary of water yield experiments at Hubbard Brook Experimental Forest, NH. *Canadian Journal of Forest Research* 27: 2043–2052.

Knoepp, J. D., and W. T. Swank. 1997. Long-term effects of commercial sawlog harvest on soil cation concentrations. *Forest Ecology and Management* 93: 1–7.

Mattson, K. G., W. T. Swank, and J. B. Waide. 1987. Decomposition of woody debris in a regenerating, clearcut forest in the Southern Appalachians. *Canadian Journal of Forest Research* 17: 712–721.

Mulholland, P. J., R. O. Hall, D. J. Sobota, W. K. Dodds, S. E. G. Findlay, N. B. Grimm, S. K. Hamilton, W. H. McDowell, J. M. O'Brien, J. L. Tank, L. R. Ashkenas, L. W. Cooper, C. N. Dahm, S. V. Gregory, S. L. Johnson, J. L. Meyer, B. J. Peterson, G. C. Poole, H. M. Valett, J. R. Webster, C. P. Arango, J. J. Beaulieu, M. J. Bernot, A. J. Burgin, C. L. Crenshaw, A. M. Helton, L. T. Johnson, B. R. Niederlehner, J. D. Potter, R. W. Sheibley, and S. M. Thomas. 2009. Nitrate removal in stream ecosystems measured by ^{15}N addition experiments: denitrification. *Limnology and Oceanography* 54: 666–680.

Patric, J. H., J. O. Evans, and J. I. Helvey. 1984. Summary of sediment yield data from forested land in the United States. *Journal of Forestry* 82: 101–104.

Potter, C. S. 1992. Stemflow nutrient inputs to soil in a successional hardwood forest. *Plant and Soil* 140: 249–254.

Potter, C. S., H. L. Ragsdale, and W. T. Swank. 1991. Atmospheric decomposition and foliar leaching in a regenerating Southern Appalachian forest canopy. *Journal of Ecology* 79: 97–115.

Swank, W. T., and W. H. Caskey. 1982. Nitrate depletion in a second-order mountain stream. *Journal of Environmental Quality* 11: 581–584.

Swank, W. T., and D. A. Crossley, Jr. 1988. Introduction and site description. Pages 3–16 in *Forest Hydrology and Ecology at Coweeta*. W. T. Swank and D. A. Crossley, Jr., editors. Springer-Verlag, New York, New York.

Swank, W. T., J. E. Douglass, and G. B. Cunningham. 1982. Changes in water yield and storm hydrographs following commercial clearcutting on a southern Appalachian catchment. Pages 583–594 in *Hydrological Research Basins and their use in Water Resource Planning*. Swiss National Hydrological Service Special Publication, Berne, Switzerland.

Swank, W. T., L. J. Reynolds, and J. M. Vose. 1992. Annual mean atmospheric deposition, through stemflow, and soil solution fluxes for major ions at Coweeta hardwood site. Page 634 in *Atmospheric Deposition and Forest Nutrient Cycling*, D. W. Johnson and S. E. Lindberg, editors. Springer-Verlag, New York, New York.

Swank, W. T., L. W. Swift, Jr., and J. E. Douglass. 1988. Streamflow changes associated with forest cutting, species conversions, and natural disturbances. Pages 297–312 in *Forest Hydrology and Ecology at Coweeta*. W. T. Swank and D. A. Crossley, Jr., editors. Springer-Verlag, New York, New York.

Swank, W. T., and J. M. Vose. 1988. Effects of cutting practices on microenvironment in relation to hardwood regeneration. Pages 71–88 in *Guidelines for Regenerating Appalachian Hardwood Stands*. H. C. Smith, A. W. Perkey, and W. E. Kidd, Jr., editors. SAF Publ. 88-03. West Virginia University Books, Morgantown.

Swank, W. T., J. M. Vose, and K. J. Elliott. 2001. Long-term hydrologic and water quality responses following commercial clearcutting of mixed hardwoods on a southern Appalachian catchment. *Forest Ecology and Management* 143: 163–178.

Swank, W. T., and J. B. Waide. 1988. Stream chemistry responses to disturbance. Pages 339–357 in *Forest Hydrology and Ecology at Coweeta*. W. T. Swank and D. A. Crossley, Jr., editors. Springer-Verlag, New York, New York.

Swift, L. W., Jr. 1972. Effect of forest cover and mountain physiography on the radiant energy balance. Dissertation. Duke University, Durham, North Carolina.

Swift, L. W., Jr. 1988. Forest access roads: design, maintenance, and soil loss. Pages 313–324 in *Forest Hydrology and Ecology at Coweeta*. W. T. Swank and D. A. Crossley, Jr., editors. Springer-Verlag, New York, New York.

Swift, L. W., Jr., G. B. Cunningham, and J. E. Douglass. 1988. Climatology and hydrology. Pages 35–55 in *Forest Hydrology and Ecology at Coweeta*. W. T. Swank and D. A. Crossley, Jr., editors. Springer-Verlag, New York, New York.

Swift, L. W., Jr., and W. T. Swank. 1981. Long term responses of streamflow following clearcutting and regrowth. *Hydrological Sciences Bulletin* 26: 245–256.

Waide, J. B., W. H. Caskey, R. L. Todd, and L. R. Boring. 1988. Changes in soil nitrogen pools and transformations following forest clearcutting. Pages 221–232 in *Forest Hydrology and Ecology at Coweeta*. W. T. Swank and D. A. Crossley, Jr., editors. Springer-Verlag, New York, New York.

4

Long- and Short-Term Changes in Nutrient Availability Following Commercial Sawlog Harvest via Cable Logging

Jennifer D. Knoepp*
Wayne T. Swank
Bruce L. Haines

Introduction

Soil nutrient availability often limits forest productivity, and soils vary considerably in their ability to supply nutrients (Cole 1995). Most southern Appalachian forests are minimally managed, with no fertilizer inputs or routine thinning regime. Nutrient availability is regulated by atmospheric inputs and the internal cycling of nutrients through such processes as forest floor decomposition (Hursh 1928; Alban 1982), soil organic matter mineralization, and the weathering of parent material or primary minerals. Long-term studies in undisturbed forests have shown that soil cation concentrations often decline over time (Binkley et al. 1989; Billett et al. 1990; Knoepp and Swank 1994; Richter et al. 1994). These decreases result from the sequestration of nutrients in aboveground biomass and/or leaching to streams (Johnson et al. 1988; Knoepp and Swank 1994). The sequestration of nutrients has raised questions about the long-term effects of harvesting vegetation and its associated nutrient removal on long-term soil nutrient availability and site productivity.

The responses of soil nutrient pools to forest harvesting vary according to harvest method and forest type. Research has shown that a whole-tree harvest can decrease both total N and exchangeable base cations (Mroz et al. 1985; Waide et al. 1988; Knoepp and Swank 1997). While nutrient declines do not always occur (Hendrickson et al. 1989), the potential for such losses has raised concerns about long-term forest productivity when whole-tree harvesting methods are

* Corresponding author: Coweeta Hydrologic Laboratory, USDA Forest Service, 3160 Coweeta Lab Road, Otto, NC 28763 USA

repeated over several rotations (Federer et al. 1989). In contrast, studies show that a commercial sawlog harvest either increases or has no effect on soil nutrient concentrations (Hendrickson et al. 1989; Kraske and Fernandez 1993), suggesting no negative impact on long-term site productivity (Waide et al. 1988).

Several authors have reported short-lived increases in soil cation concentrations following sawlog harvest (Snyder and Harter 1984; Hendrickson et al. 1989; Kraske and Fernandez 1993). However, forest soil total C and N show varied responses to harvesting depending on site history and treatment (Knoepp and Swank 1997; Johnson et al. 2002b; Johnson et al. 2007). Knoepp and Swank (1997) demonstrated the importance of harvest slash on soil total C and N responses, comparing a whole-tree harvest site with no soil C changes and decreased soil N and a stem-only harvest in which both soil C and N increased.

Forest clearcutting results in changes in nutrient inputs, and nutrient movement within a watershed may occur. While wet deposition (precipitation) is unchanged by forest cutting, dry deposition inputs decrease due to the removal or alteration in aboveground surface area (Boring et al. 1988b). Swank and Henderson (1976) found that dry deposition contributed 19% to 31% of the annual cation input and 7% to 12% of anion input at Coweeta. Total dry deposition was lower by up to 50% in Coweeta than in similar forest communities in the Walker Branch Watershed in Oak Ridge, Tennessee. This difference was attributed to the proximity of coal power plants in Tennessee. Forest structure also has a significant effect on dry deposition input to forests. Studies of cloud and dry deposition to forests and forest edges conducted in the Great Smoky Mountains National Park showed that trees growing on the edge of the forest received about 3 times the input of forest saplings within a gap (Lindberg and Owens 1993). This suggests that inputs may decrease substantially following removal of the forest canopy. Data collected by Potter et al. (1991) in 1985, eight years following site harvest, found that dry-deposition estimates were 2 to 5 times lower than in the Walker Branch Watershed. While some of this difference was attributed to total dry deposition, the effect of the forest canopy was also discussed. In 1985, the WS 7 canopy was estimated to have 85% leaf area index (LAI) of a mature forest, therefore decreasing the rate of dry deposition. Little is known about the recovery of this input source to forests following harvest.

Increases in N availability and increased rates of N transformation processes, mineralization and nitrification, following a forest disturbance or harvest have been reported (Waide et al. 1988; Donaldson and Henderson 1990; Smethurst and Nambiar 1990; O'Connell et al. 2004; Lapointe et al. 2005). It has been suggested that the response is related to the intensity of disturbance, which results in increased soil temperature and water content. Smethurst and Nambiar (1990) found that clearcutting resulted in increased rates of N mineralization in the litter layer 4 months following site disturbance and in the mineral soil after 1 year. This led to increased soil solution concentrations of inorganic N, which leached to depths greater than 30 cm, which is below the major rooting zone. Inorganic N concentrations remained elevated for at least 3 years. Increases in soil N availability were greater in sites where logging slash was left on site. Nutrient movement and losses are related both to the disturbance response and changes in water movement through the soil profile (Mann et al. 1988; Swank and Vose 1988; Tiedemann et al. 1988; Dahlgren and Driscoll 1994).

Studies showing the effects of harvest on soil nutrients often rely on either short-term data or the chronosequence approach to estimate long-term responses (Mroz et al. 1985; Mann et al. 1988; Hendrickson et al. 1989; Kraske and Fernandez 1993). The objective of our study was to examine the changes in nutrient availability in WS 7 immediately following and for 22 years following harvest. We measured several indices of nutrient availability and nutrient cycling rates including total soil C and N, soil extractable cations, extractable NO_3 and NH_4, net soil N transformations, soil solution NO_3 concentrations, and nutrient fluxes in slash throughfall. Responses to cutting were measured against both pretreatment soil conditions and by comparison with data collected from an adjacent reference watershed.

Materials and Methods

Site Description

This study was conducted at the USDA Forest Service Coweeta Hydrologic Laboratory in western North Carolina. The clearcut watershed (WS 7) is 59 ha, has slopes of 23% to 81%, and ranges in elevation from 722 to 1077 m. The site was harvested in 1976–1977 using a cable-yarding system. The 12-ha reference watershed (WS 2) has remained untreated since 1927 and ranges in elevation from 709 to 1004 m, with slopes similar to WS 7.

Soil Description

The soils found on the harvested and reference watersheds are similar; fine-loamy to coarse-loamy in texture derived from material weathered from high-grade metamorphosed mica-rich rock or from colluvium. Side-slope soils are in the Fannin and Chandler series and range from 15% to 95% slope. They are well-drained to excessively drained, very deep (solum thickness > 1 m), and > 1.8 m to bedrock. The saprolite layer beneath the solum may be up to 6 m deep (Thomas 1996). Cove or stream side soils were formed in colluvium, have 15% to 50% slope, and are mapped in the Cullasaja-Tuckasegee complex. These soils are very deep and well drained, with solum thickness < 1.5 m and depth to bedrock > 1.8 m (Thomas 1996).

WS 7 sample plots were located on Chandler series (coarse-loamy, micaceous, mesic Typic Dystrochrepts); Fannin series (fine-loamy, micaceous, mesic Typic Hapludults); Cullasaja-Tuckasegee complex (loamy-skeletal or coarse-loamy, mixed, mesic Typic Haplumbrepts); and rock outcrop–Cleveland complex (loamy, mixed, mesic Lithic Dystrochrepts). WS 2 soils include the Fannin series (side-slope) and Cullasaja-Tuckasegee complex (streamside) similar to WS 7. All sample plots were located on the Fannin soil type, which occupies about 60% of the watershed.

Sample Collection and Analysis

WS 7 sample plots ($100\ \text{m}^2$) were established in 1975 at randomly selected points along 4 transects crossing the watershed (Waide et al. 1988). Pretreatment sampling began in 1975 on 16 plots divided into 2 groups of 8; each group of 8 was sampled alternately every 2 weeks for 17 months. Posttreatment samples were collected on 10 plots, again divided into 2 groups, with alternate groups sampled every 2 weeks. Plots selected for posttreatment soil sampling included both pretreatment plots and sites of intensive vegetation inventory and tree N fixation studies. The biweekly sampling interval continued for 17 months after completion of the harvest. Subsequently, collection frequency decreased, but the alternate-group sampling design continued until 1985. We resampled all 10 plots in 1992, 1993, 1994, 1998, and 2008.

Four $100\ \text{m}^2$ sample plots on WS 2 were established following the WS 7 harvest in 1977 to serve as long-term reference plots. Plots were divided into 2 groups of 2 and sampled alternately every 2 weeks beginning in July 1977 coinciding with sample collection on WS 7.

Soil Chemical Analyses

Soils in both watersheds were sampled at two depths, 0–10 cm and 10–30 cm. Soils were air-dried and sieved to < 2 mm. Soil C determinations before 1990 were made using the Walkley-Black method (Nelson and Sommers 1982). Regression analysis showed a good correlation between Walkley-Black (WB) and the Leco carbon analyzer with a slope of 1.01 and $r^2 = 0.99$ (n = 24). After 1990, soil C was determined using combustion analysis on a Perkin-Elmer 2400 CHN analyzer. Comparisons of total percent C determined with the Leco and the Perkin-Elmer 2400 were conducted using an internal soil standard (4.8% C), certified standard $CaHCO_3$, and certified standard acetanilide. No significant differences in slope or intercept were detected.

Total N concentrations for all pre-1990 samples were determined using the micro-Kjeldahl method (total Kjeldahl Nitrogen, TKN). Digested solution was analyzed using the cyanurate-salicylate reaction with a segmented flow autoanalyzer (Bremner and Mulvaney 1982). Soil samples collected between 1992 and 2008 were analyzed for total percent N by the combustion method using a Perkin-Elmer 2400 CHN analyzer. Regression analysis between percent N as determined with TKN and combustion was conducted to insure appropriate data comparison. Total percent N by combustion was determined on archived soil samples collected and analyzed for TKN in 1982. Analysis showed a good relationship between the TKN values obtained in 1982 and combustion data with a slope of 1.08 and $r^2 = 0.98$ (n = 24). Quality control during soil analysis after 1990 was conducted for total N and C on each set of soil samples.

Soil cation concentrations were determined with the dilute double acid extraction procedure developed by the Soil Testing and Plant Analyses Laboratory, Cooperative Extension Service, Athens, Georgia. Five-gram soil samples were

Table 4.1 Regression analysis output for plasma emission spectroscopy versus atomic absorption spectroscopy analysis of double acid soil extracts.

	Ca^{2+}	Mg^{2+}	K^+
Slope	0.95	0.93	0.94
P-value	(< 0.001)	(< 0.001)	(< 0.001)
Intercept	32.2	2.16	2.50
P-value	(0.07)	(0.003)	(0.07)
r^2	0.99	0.99	0.99

Note: Data analyzed were 1993 WS 7 soil cation concentrations (n = 24). Regression values shown with probability of significant value.

extracted with 20 ml of 0.05 M HCl plus 0.05 M H_2SO_4. Pre-1990 soil cation determinations were made using plasma emission spectroscopy (PES) as described by Jones (1977). Although extraction methods did not change in the 1990s, atomic absorption spectroscopy (AAS) was used for base cation concentration determinations. Regression analysis on the 1993 samples compared PES with AAS (table 4.1). AAS data were transformed using these equations before statistical analysis.

Soil Nitrogen Availability Determination

Soil NO_3 and NH_4 concentrations were measured on most soil samples collected for chemical analysis as described above from 1975 to 1980 for WS 7 and 1977 to 1982 for WS 2. We measured soil NO_3 and NH_4 concentrations on all plots in spring 1998 and summer 1999. We conducted laboratory soil incubations (0–10 cm) to determine rates of N mineralization and nitrification during the 1999 growing season in both WS 7 and WS 2 using the methods outlined by Waide et al. (1988). Samples were collected using 15-cm-long, 4.3-cm inside diameter PVC core. Two cores were driven 10 cm into the ground at random locations in each permanent plot. Soil from each core was composited into a single sample for determining NO_3 and NH_4 concentrations and conducting laboratory incubations for potential N mineralization rates. Initial NO_3 and NH_4 concentrations were determined in triplicate for each composited soil sample by extracting 5 g of fresh soil with 20 ml of 2 M KCl. Soil moisture content was determined by drying a 10-g sample overnight at 105°C. We weighed three 10-g subsamples of each composite into 0.94-L jars for laboratory incubation. Jars plus soil were placed in a 25°C incubator (time = 0); soil moisture was adjusted to 33% the following day after soil moisture determinations were complete. Jars were covered with plastic wrap and soil moisture was adjusted weekly as necessary. After 33 days of incubation, 40 ml of 2 M KCl was added to each jar plus soil to extract NO_3 and NH_4. Nitrification rates equal NO_3-N concentration at 33 days minus NO_3-N at time zero. Nitrogen mineralization rates equal NH_4-N + NO_3-N at 33 days minus NH_4-N + NO_3-N at time zero.

Slash Throughfall

We established two, 4 x 4 m plots in each of nine locations (18 plots) in WS 7, which were stratified to proportionally represent precutting forest type for a wood decomposition study. Wood sampling took place 6, 7, and 11 years after cutting (Mattson et al. 1987; see also Mattson and Swank, chapter 7, this volume). Within each 4 x 4 m plot, one nested 2 x 2 m plot was sampled for determinations of fine wood (branches and stems < 5 cm in diameter) inputs following logging and site preparation operations. Plots were selected to represent the range of slash dominated by stems, brush, or mixed slash, all < 5 cm in diameter. Coarse wood was subsampled to estimate surface area and biomass.

We measured slash throughfall by inserting a 15 x 200 cm V-shaped aluminum trough beneath the slash, attached to 19-L polypropylene collection jugs with tygon tubing (figure 4.1). Samples were collected on a storm event basis and used to estimate both slash rainfall interception and nutrient leaching. Samples for each trough were collected weekly and composited monthly by volume yielding weighted composite samples. Solution analyses for NH_4 and NO_3 were conducted on monthly composites as previously described; TKN analyses were conducted on samples composited quarterly during the calendar year; total organic nitrogen (TON) was calculated as TON = TKN-(NO_3-N + NH_4-N). We used the annual average throughfall TON concentration and the

Figure 4.1 Throughfall collector beneath slash (one of 12 plots), March 1978. (USDA Forest Service photo)

total throughfall volume to calculate TON fluxes through the slash layer. Data presented are fluxes in g ha^{-1} y^{-1}.

Soil Solution Collection

After harvest soil solution was collected using both continuous tension porous plate and falling tension 5-cm-diameter porous cup lysimeters. All lysimeters were washed prior to field installation using 1 N HCl (Grover and Lamborn 1970) followed by rinsing with deionized water until electrical conductivity of the leachate equaled the deionized water. Porous plate lysimeters were constructed from 15-cm-diameter ceramic plates (Pacific Lysimeter, Seattle, Washington) fitted with a fiberglass resin backing and plastic drain tubes. Solutions drained from the lysimeters to 25-L vented plastic carboys. Tension porous plate lysimeters were inserted 30 cm below the soil surface. Drain elevations were 100 cm below the surface of the ceramic plates, and were filled with water providing a 100-cm hanging water column, approximately −0.01 MPa. Porous cup lysimeters were installed in pairs at 30 and 100 cm beneath the litter-soil interface. Porous cup lysimeters were evacuated weekly with a hand pump to −0.03 MPa.

Tension porous plates were placed at 30 cm depth in all 10 plots on WS 7 and 4 on WS 2. Samples were collected once a month for analysis during 1978 and 1979. Falling tension porous cup lysimeters were installed at 30 cm and 100 cm and sampled monthly 1980 through 1984. Porous cup lysimeters were installed on a subset of the sample plots with similar vegetation and soil types, four on WS 7 and three on WS 2. Soil solution analyses were conducted as described for stream water by Swank et al. (see chapter 3, this volume).

Statistical Analysis

Statistical analyses for differences between pretreatment and posttreatment years in soil extractable Ca, Mg, and K concentrations for each watershed were tested through analysis of variance with the Mixed Procedure of SAS (SAS 2000). We used WS(Plot) in the random statement and determined significant differences ($P \leq 0.10$) among means using Tukey adjusted LSMeans. All WS 7 pretreatment soil-chemistry data from 1975 and 1976 were compared with annual posttreatment means. We compared WS 7 and WS 2 soil chemistry and soil solution chemistry year-by-year using the 6 plots on WS 7 with soils similar to those on WS 2. We conducted analysis of variance with the GLM Procedure of SAS (SAS 2000) using WS(Plot) as the error term, with significant differences ($P \leq 0.10$) reported. Plot soil series determination was based on the Coweeta soil map (Thomas 1996) with plot level ground truthing.

Differences between WS 7 and WS 2 soil NO_3 and NH_4 concentrations (t = 0) for all sample years were determined as described above using annual plot means from plots in similar soil types on WS 7 and WS 2. Pre-1999 data for laboratory incubation N mineralization and nitrification rate measurements were not available;

Table 4.2 Precipitation inputs and net throughfall inputs for WS 7 postharvest logging slash and WS 2.

Year	Site/Material	NO_3-N	NH_4-N	PO_4	SO_4	Ca	K
Precipitation inputs (g ha^{-1} yr^{-1})							
1977-1979		2680	1569	251	28674	3950	1928
Net throughfall (g ha^{-1} yr^{-1})							
1978	WS 7 postharvest slash	1251	694	−3843	−25610	−9701	−35929
1985–1988[a]	WS 2 reference forest	1351	−826	−325	−2928	−7240	−21489

Note: Net throughfall equals precipitation input minus throughfall. Values shown are annual means for the time period. Net throughfall values < 0 indicate nutrient leaching from canopy and values > 0 indicate canopy uptake.
[a]Throughfall data collected during the Integrated Forest Study (Swank et al. 1992).

therefore, statistical comparisons between past and recent data were not possible. Previously published mean values and standard errors are presented (Waide et al. 1988).

Results

Nutrient Input Response

Post-harvest total bulk precipitation of nutrients was typical compared to years before forest harvest (table 4.2). However, the removal of the overstory changed nutrient input to the site. For example, slash throughfall chemistry on WS 7 varied substantially in NH_4, SO_4, and PO_4 compared to throughfall concentrations in the adjacent reference watershed, WS 2, measured during the Integrated Forest Study funded by the Electric Power Research Institute (Swank et al. 1992) resulting in differences in net throughfall (table 4.2).

Soil Carbon and Nitrogen Response

Surface soil (0–10 cm) C increased significantly after forest harvest. C content in surface soils averaged 32.1 Mg/ha during the 2 sampling years prior to harvest and increased to an average of 50 Mg/ha for 3 years after forest cutting (figure 4.2A). Surface soil C remained greater, although differences were significant in only 2 of the next 28 years of sampling. Responses of subsurface (10–30 cm) soils C differed (figure 4.2B). Total C did not respond in the first 3 years following cutting, however, beginning in 1980, subsurface C was significantly lower compared to pretreatment levels in 5 of the 13 posttreatment sampling years.

Total N in the surface soil (0–10 cm) increased significantly compared to preharvest N content for 3 years following forest harvest (figure 4.3A). Total N content increased from 1.4 Mg/ha preharvest mean to 2.2 Mg/ha in 1977 through 1979. After this initial response, total N did not differ significantly from preharvest levels

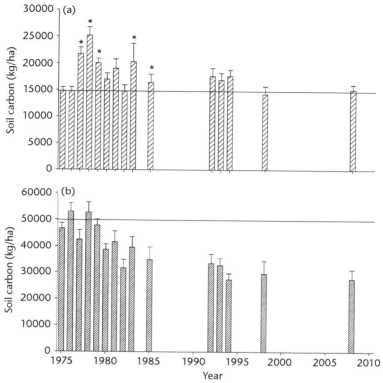

Figure 4.2   Soil carbon (C) content (kg/ha) for (**A**) 0–10 cm and (**B**) 10–30 cm collected between 1975 and 1998 on harvested WS 7. Annual means were calculated using annual mean values for each plot (preharvest, n = 16; postharvest n = 10). The line represents the mean C content of preharvest soils. Error bars represent standard errors of the mean. Bars marked with '*' have a significant difference between pre- and postharvest C content ($P \leq 0.10$).

for any of the next 8 sample collections over 19 years. Subsoil total N content averaged 3.1 Mg/ha preharvest and did not change immediately following cutting. However, subsoil N decreased to < 2.0 Mg/ha by the final sample collection in 2008 (figure 4.3B).

Comparing the seven plots on WS 7 with soils similar to WS 2, the reference watershed (postharvest collections only), suggests a different pattern of total soil C and N response over time (figs. 4.4 and 4.5). For example, surface soil total C increases were significant for 5 years following harvest and were greater in two of the next seven sample collections (figure 4.4A). On the other hand, comparing WS 7 with WS 2 suggests no significant change in total C in the subsoil layer. Comparing WS 7 and WS 2 suggests that both watersheds experienced a long-term decline in subsurface soil C during the postharvest sampling years (figure 4.4B). A comparison of WS 7 and WS 2 total N content showed a significant increase for only one year after harvest (1979, figure 4.5A) in the surface and subsurface soil layer (figure 4.5B).

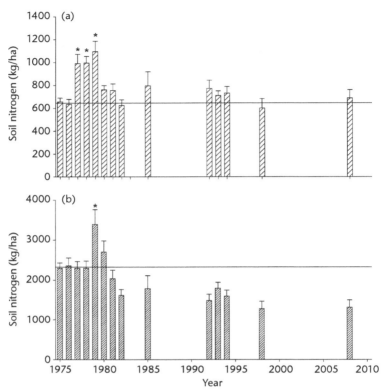

Figure 4.3 Soil nitrogen (N) content (kg/ha) for (**A**) 0–10 cm and (**B**) 10–30 cm collected between 1975 and 1998 on harvested WS 7. Annual means were calculated using annual mean values for each plot (preharvest, n = 16; postharvest n = 10). The line represents the mean N content of preharvest soils. Error bars represent standard errors of the mean. Bars marked with '*' have a significant difference between pre- and postharvest N content (*P* ≤ 0.10).

Soil Cation Responses

We measured a significant increase in surface soil Ca, Mg, and K content for 3 years (1977–1979) following harvest compared to preharvest cation content (figures 4.6A, 4.7A, 4.8A). Increases averaged over 100% for each cation. Following the initial cation increase, responses during the years 1980–2008 differed for each cation. During the post-1980 years, Ca and K content did not differ from preharvest levels. Mg content postharvest was significantly greater than it was pretreatment in 1992 and 2008.

Subsurface soils responses also differed among the three base cations. There were no significant differences between pre- and postharvest subsoil Ca content (figure 4.6B). Subsoil Mg responses were similar to surface soils, 2 of 3 postharvest years and 2 of 4 years in the 1990s had significantly greater Mg content compared to preharvest values (figure 4.7B). Soil K content response in the subsurface soil layer

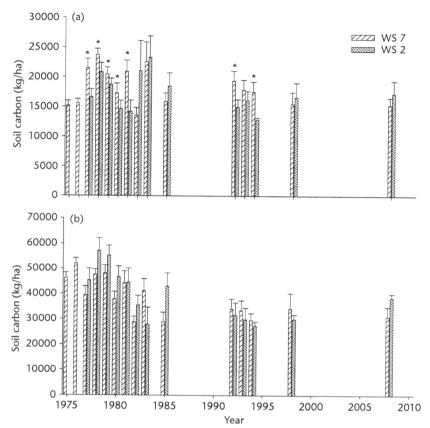

Figure 4.4   Soil C content (kg/ha) for (**A**) 0–10 cm and (**B**) 10–30 cm collected between 1977 and 1998 on harvested WS 7 and reference WS 2. Annual means were calculated using annual mean values for each plot (WS 7 n = 6; WS 2 n = 4). Error bars represent standard errors of the mean. Bars marked with '*' have a significant difference between WS 7 and WS 2 ($P \leq 0.10$).

differed from K surface soils and other cations in subsurface soils. Post-harvest K content was significantly greater than preharvest content in 10 out of 11 postharvest sample collection years (figure 4.8B).

Similar to soil C and N responses, comparing plots on harvested and reference watershed with the same soil series resulted in different conclusions regarding soil cation responses to harvest. Surface soil Ca content showed no significant response to harvest, except in 1992 (figure 4.9). However, this analysis suggests that the Ca content of the subsurface soil 10–30 cm soil layer was significantly lower in WS 7 compared to WS 2 in 3 of the first 4 postharvest years. Soil Mg content responses also differed using this comparison (figure 4.10). In 1979, 3 years after harvest, surface and subsurface soil Mg content was greater in WS 2 compared to WS 7. Soil Mg content did not differ significantly between the two watersheds in any other

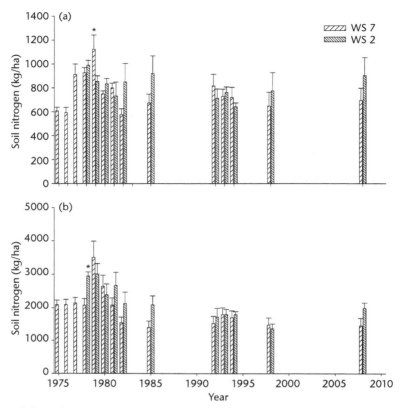

Figure 4.5   Soil N content (kg/ha) for (**A**) 0–10 cm and (**B**) 10–30 cm collected between 1977 and 1998 on harvested WS 7 and reference WS 2. Annual means were calculated using annual mean values for each plot (WS 7 n = 6; WS 2 n = 4). Error bars represent standard errors of the mean. Bars marked with '*' have a significant difference between WS 7 and WS 2 ($P \leq 0.10$).

sample years. Soil K responses were similar to the other cations (figure 4.11). There were few significant differences between WS 7 and WS 2, and in those years K content was greater in WS 2.

Nitrogen Availability

Waide et al. (1988) found significant increases in extractable soil NH_4 concentrations following harvest from 1977–1980. NH_4 concentrations averaged 8.9 mg N/kg in WS 7 compared to 3.1 mg N/kg in WS 2 (table 4.3). Differences in NO_3 concentrations between WS 7 and WS 2 were similar in magnitude, but due to high variability were not significant. The NH_4 concentrations in soils collected in 1998 and 1999 were 50% greater in WS 7 compared to WS 2, while NO_3 was 10 times greater; neither was significantly different (table 4.3).

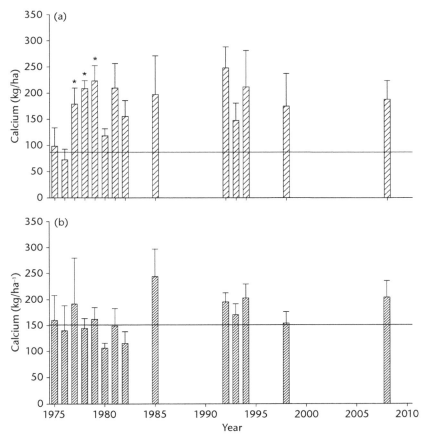

Figure 4.6 Soil calcium (Ca) content (kg/ha) for (**A**) 0–10 cm and (**B**) 10–30 cm collected between 1975 and 1998 on harvested WS 7. Annual means were calculated using annual mean values for each plot (preharvest, n = 16; postharvest n = 10). The line represents the mean Ca content of preharvest soils. Error bars represent standard errors of the mean. Bars marked with '*' have a significant difference between pre- and postharvest Ca content ($P \leq 0.10$).

Potential net N mineralization and nitrification was measured using laboratory incubations at constant temperature and soil moisture during the years 1979–1982 and again in 1999. Early data suggested a slight increase in N mineralization that could have contributed to the increase in available soil N concentrations. There was also a significant increase in nitrification rates at all soil layers tested, Oa, 0–10 cm, and 10–30 cm soil depths. Again, Waide et al. (1988) suggested that the increase in nitrification potential was consistent with the increase in soil extractable NO_3. In 1999 laboratory incubations, net N mineralization and nitrification were 3.4 and 20 times greater in WS 7 compared to WS 2, respectively. However, due to high variability these differences were not statistically significant (table 4.3).

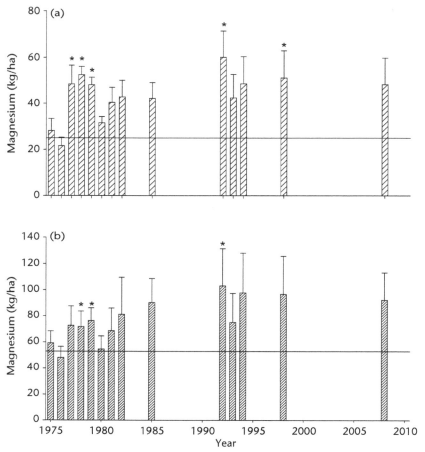

Figure 4.7 Soil magnesium (Mg) content (kg/ha) for (**A**) 0–10 cm and (**B**) 10–30 cm collected between 1975 and 1998 on harvested WS 7. Annual means were calculated using annual mean values for each plot (preharvest, n = 16; postharvest n = 10). The line represents the mean Mg content of preharvest soils. Error bars represent standard errors of the mean. Bars marked with '*' have a significant difference between pre- and postharvest Mg content ($P \leq 0.10$).

Soil Solution Responses

Postharvest annual means of soil solution collected using tension lysimeters at 30 cm showed that overall WS 7 had significantly greater NO_3 and Ca in solution compared to WS 2 (figure 4.12). In yearly comparisons of WS 7 and WS 2 soil solution concentrations of NO_3 in 1984, Ca in 1983 and 1984, and K in 1984 were significantly greater in WS 7. In 1983, Ca concentrations in soil solutions collected at 100 cm were also significantly greater in WS 7 compared to WS 2 (figure 4.13).

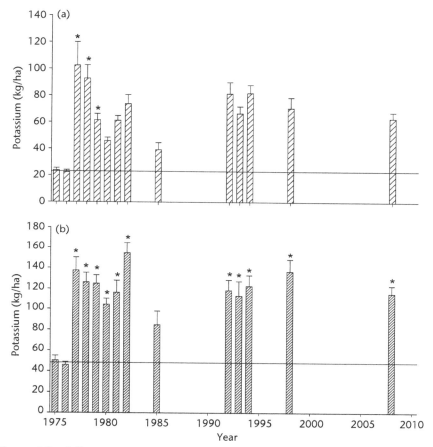

Figure 4.8   Soil potassium (K) content (kg/ha) for (**A**) 0–10 cm and (**B**) 10–30 cm collected between 1975 and 1998 on harvested WS 7. Annual means were calculated using annual mean values for each plot (preharvest, n = 16; postharvest n = 10). The line represents the mean K content of preharvest soils. Error bars represent standard errors of the mean. Bars marked with '*' have a significant difference between pre- and postharvest K content ($P \leq 0.10$).

Discussion

Nutrient Inputs

Patterns of throughfall collected beneath logging slash compared to undisturbed forest are indicative of the impacts of forest harvest on nutrient inputs via dry deposition and impacts of canopy processes on nutrient fluxes (table 4.2). The net input of NH_4 to the forest floor was lower in the clearcut forest compared to the reference; the slash canopy immobilized N, while the N was removed from the reference canopy. Inputs of PO_4 and SO_4 to the forest floor were also altered, and while both were leached from slash and canopy materials, the leaching of PO_4 and SO_4 was

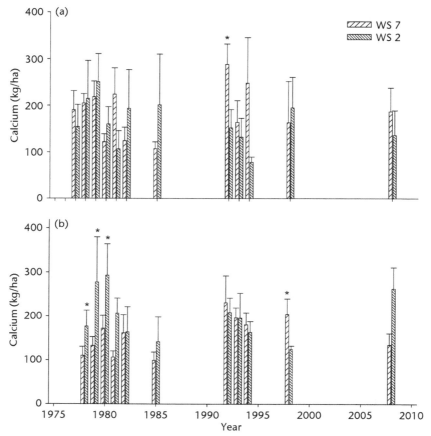

Figure 4.9 Soil Ca content (kg/ha) for (**A**) 0–10 cm and (**B**) 10–30 cm collected between 1977 and 1998 on harvested WS 7 and reference WS 2. Annual means were calculated using annual mean values for each plot (WS 7 n = 6; WS 2 n = 4). Error bars represent standard errors of the mean. Bars marked with '*' have a significant difference between WS 7 and WS 2 ($P \le 0.10$).

greater from the slash material. As the forest regrows and the canopy structure and LAI return to preharvest conditions, these values should recover. A detailed study of atmospheric deposition and foliage leaching was conducted on WS 7 when the regenerating forest was 8 years old (Potter et al. 1991; Potter 1992). They collected rainfall and dry deposition as well as throughfall and stemflow in an area dominated by chestnut oak (*Quercus prinus*). These water fluxes for the regenerating stand were similar in magnitude to those measured concurrently in the adjacent reference hardwood forest on WS 2 (Swank et al. 1992).

Forests of the southern Appalachians are rarely fertilized and so nutrient availability is regulated by inputs to the system from the atmosphere and soils plus the

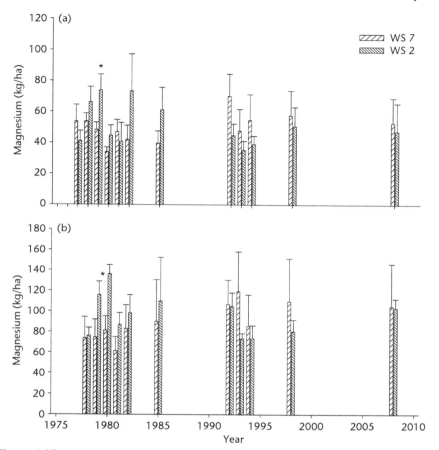

Figure 4.10. Soil Mg content (kg/ha) for (**A**) 0–10 cm and (**B**) 10–30 cm collected between 1977 and 1998 on harvested WS 7 and reference WS 2. Annual means were calculated using annual mean values for each plot (WS 7 n = 6; WS 2 n = 4). Error bars represent standard errors of the mean. Bars marked with '*' have a significant difference between WS 7 and WS 2 ($P \leq 0.10$).

cycling of nutrients input through litterfall and subsequent decomposition (Hursh 1928; Alban 1982). Research examining inputs by throughfall and litterfall by Eaton et al. (1973) and later by Gallardo et al. (1998) suggested that C, N, and P associated with organic compounds move as litterfall from the canopy to the forest floor and then to the soil through decomposition; whereas ions such as K move rapidly from the canopy to the soil by throughfall and stemflow. Patterns of canopy retention or leaching differ by nutrient both seasonally and spatially (Eaton et al. 1973; Andersson 1991; Knoepp et al. 2008; Knoepp et al. 2011).

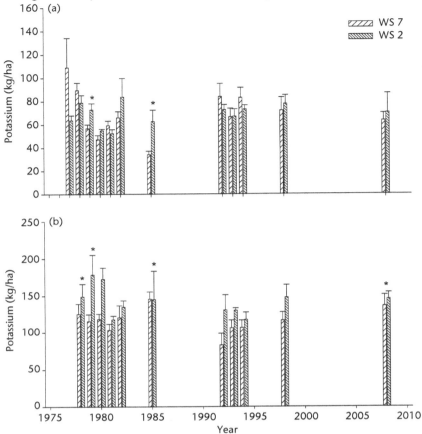

Figure 4.11. Soil K content (kg/ha) for (**A**) 0–10 cm and (**B**) 10–30 cm collected between 1977 and 1998 on harvested WS 7 and reference WS 2. Annual means were calculated using annual mean values for each plot (WS 7 n = 6; WS 2 n = 4). Error bars represent standard errors of the mean. Bars marked with '*' have a significant difference between WS 7 and WS 2 ($P \leq 0.10$).

Nutrient Pools

We measured an increase in total soil C and N in the sawlog-only harvest in this study; however, soils in a whole-tree harvest site had no C response and a significant decrease in N following cutting (Knoepp and Swank 1997). Many research studies attribute the initial soil nutrient responses to inputs of slash or logging residue to the forest floor. For example, Shammas et al. (2003) examined the impacts of harvest slash on nutrient cycling in Eucalyptus plantations in Australia. They estimated that the total N in the logging residue equaled 11% of the total N in the surface soils, but up to 100% of exchangeable cations. The largest pool of nutrients was in the leaves, which contributed 30% of the total mass and 75% of the total N. In a *Pinus radiata* plantation in Australia, Baker et al. (1989) examined

Table 4.3 Soil N transformation rates and KCl extractable N concentrations for the surface soils (0–10 cm) on WS 7 and WS 2.

A.	1979–1982		1999		
	WS 2	WS 7	WS 2	WS 7	WS 7 (n = 6)
Mineralization	17.8	21.7	5.8	25.5	17.0
(mg N kg^{-1} 33d^{-1})	(3.2)	(3.1)	(1.4)	(6.5)	(5.7)
Nitrification	4.5	15.4	0.3	6.3	2.6
(mg N kg^{-1} 33d^{-1})	(1.7)	(3.1)	(0.2)	(1.9)	(1.4)

B.	1977–1980		1998–1999		
	WS 2	WS 7	WS 2	WS 7	WS 7 (n = 6)
NH$_4$-N	3.1	8.9	1.8	2.7	2.04
(mg/kg)	(0.3)	(1.2)	(0.41)	(0.41)	(0.28)
NO$_3$-N	1.6	3.2	0.01	1.0	0.17
(mg/kg)	(0.4)	(0.6)	(< 0.01)	(0.47)	(0.10)

Note: (**A**) Data presented for 1979–1982 are from Waide et al. (1988). The 1999 data are means of 2 laboratory incubations (25°C) conducted in May and July (standard errors in parentheses). Mineralization and nitrification rates presented are for all WS 7 plots (comparable to Waide et al. 1988) as well as plots (n = 6) with soils similar to those on WS 2. (**B**) Soil ammonium and nitrate nitrogen concentrations at collection (prior to incubation) for incubations conducted 1977–1980 as well as in situ incubations conducted 1998–1999. Concentrations presented are for all WS 7 plots (comparable to Waide et al. 1988) as well as plots (n = 6) with soils similar to those on WS 2.

patterns of nutrient release from logging slash. They found that nutrient leaching began after 3 months of decomposition and that after 12 months, 16% of the total N had been released, while 19% of the Ca and 92% of K had been leached from the slash material.

The cable-yarding sawlog harvest method used in this study, removed only about 10% of the total aboveground biomass (Boring et al. 1988a), leaving 121 Mg/ha of logging residue (Mattson et al. 1987) fairly well distributed across the site. Mattson et al. (1987) estimated a composition of 91.2 Mg/ha of coarse woody material and 30.3 Mg/ha fine wood. The branches, boles, and roots represent a long-lasting source of nutrients and organic matter. Inventory of woody residue 6 years after cutting found 58% of the coarse wood and 25% of the fine wood still remaining on the site (Mattson et al. 1987).

Leaves and small twigs, which decompose rapidly, provide another immediate source of exchangeable cations and soil organic matter (Abbott and Crossley 1982) and may represent an immediate source of cations to the soil beneath the harvest slash (table 4.2). Decomposition of this material initially immobilizes some nutrients, such as Ca, but subsequently releases them as it continues to decompose (Abbott and Crossley 1982; Fahey et al. 1988). This process may also limit the loss of dissolved organic material to streams that occurs following harvest (Mann et al. 1988; Swank et al. 1988; Tiedemann et al. 1988; Dahlgren and Driscoll 1994).

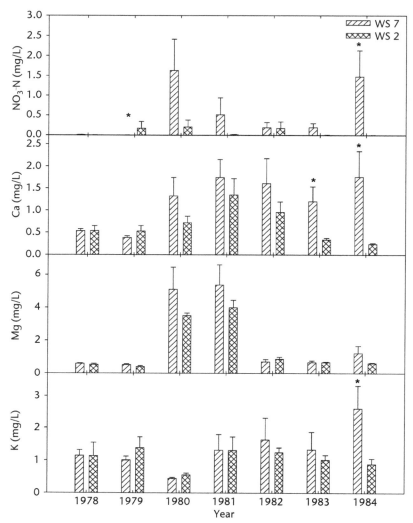

Figure 4.12 Soil solution concentrations (mg/L) of NO_3-N, Ca, Mg, and K, collected using continuous tension porous plate (1978 and 1979) and falling tension porous cup lysimeters (1980–1984) placed at a soil depth of 30 cm. Values represent means of annual plot means for WS 7 (5 plots) and WS 2 (4 plots), bars represent the standard error of the mean. Bars marked with '*' have a significant difference between WS 7 and WS 2 ($P \leq 0.10$).

Long-term Soil C and N Response

In before and after comparisons, our data show that after harvest there was an initial increase in surface soil total C and N followed by a long period without further change. However, comparison of harvested WS 7 with reference WS 2 shows a decline in total C and N over the 30+-year sampling period for both watersheds.

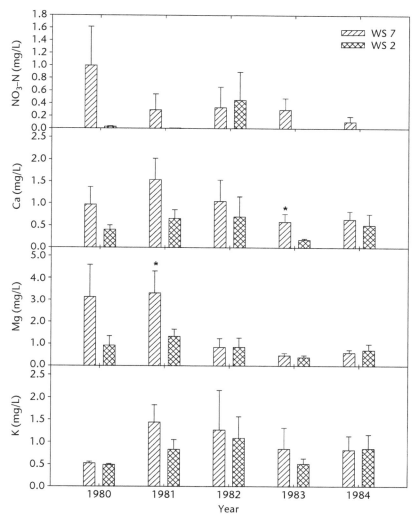

Figure 4.13 Soil solution concentrations (mg/L) of NO_3-N, Ca, Mg, and K, collected using falling tension porous cup lysimeters (1980–1984) placed at a soil depth of 100 cm. Values represent means of annual plot means for WS 7 (5 plots) and WS 2 (4 plots), bars represent the standard error of the mean. Bars marked with '*' have a significant difference between WS 7 and WS 2 ($P \leq 0.10$).

Studies of long-term changes in forest soil total C and N show varied responses depending on site history and treatment (Knoepp and Swank 1997; Johnson et al. 2002b; Johnson et al. 2007). Both Johnson et al. (2002b) and Knoepp and Swank (1997) found that long-term sampling of forested sites showed temporal variation and patterns were not always consistent in direction of change. Comparing sites that underwent change in vegetation community cover with sites that remained forested, Mueller and Kogel-Knabner (2009) found that forested sites that remained in forest had greater stocks of total C. They attributed this to the high content of C that is physically associated with soil clay particles in the forested sites, C that was therefore protected from further decomposition. Differences in total soil C have all been attributed to differences among vegetation types, site management, and site history. Paul et al. (2003) linked a C accounting model with forest growth models and a soil C model to examine the factors controlling total C change over time in forest plantations. They concluded that overall differences in total mass and quality of organic matter input to soils regulated rates of C accumulation. This supports the hypothesis that initial changes in total C following harvest are the result of slash inputs, changes in rooting density, and rates of decomposition (Fahey et al. 1988). Mattson and Swank (1989) found that 6 years after cutting, fine-root biomass on WS 7 was only 65% of that of WS 2, the adjacent reference watershed. This decrease could be the result of extensive root mortality and/or changes that occurred in vegetation composition and age after stand harvest. Changes in root density following harvest were evident in a study by Edwards and Ross-Todd (1983). They found decreased in situ soil CO_2 efflux rates in forested sites after harvest and concluded it was the result of reduced root respiration. In laboratory studies they found that respiration from soil alone was greater in soils from harvested sites (Edwards and Ross-Todd 1983). Scott et al. (2004) looked at changes in ecosystem C after harvest using eddy covariance measurements, examining a site preharvest and after removal of 30% aboveground biomass. This cutting intensity resulted in a 40% decline in litterfall and LAI, resulting in net ecosystem exchange of CO_2 decline of 18% the first year. Their data showed a slight decline in soil respiration.

Soil Nitrogen Availability and Mobility

Forest harvesting or disturbance often results in increased soil N availability and increased rates of N transformation processes, mineralization and nitrification (Waide et al. 1988; Donaldson and Henderson 1990; Smethurst and Nambiar 1990; O'Connell et al. 2004; Lapointe et al. 2005). Some of these studies have found increased response with increasing level of disturbance response. Lapointe et al. (2005) examined a chronosequence, looking at trembling aspen to late successional boreal forests and found that partial harvest resulted in less increase in N mineralization potential compared to clearcutting. In a study examining the relationships of site quality and N responses, O'Connell et al. (2004) looked at soil N responses to logging residue in short rotation Eucalyptus plantations. They found that inorganic N concentrations increased for 2 years following harvest, until new vegetation developed. At the more fertile sites, inorganic N was present mostly as NO_3 and in areas where logging residue was left on site, NO_3 concentrations were lower. The impact of residue was less pronounced at the low N site.

Our data show a continued increase in N availability in surface soils of WS 7 more than 20 years following harvest (table 4.3). We attribute this not only to the large input of logging residue as described above but also the changes in vegetation community structure that have occurred since harvest. Elliott et al. (1997) and Elliott et al. (see chapter 2, this volume) described the change in forest composition through succession following clearcutting WS 7. *Quercus* and *Carya* spp. dominated the preharvest forest, while the postharvest forest is dominated by *Liriodendron*, *Robinia*, and *Acer*. Changing vegetation following harvest can result in changes in nutrient demand (Hendrickson 1988) and water demand (see Swank et al., chapter 3, this volume). Vegetation composition also directly impacts soil nutrient content and nutrient transformation processes. Knoepp et al. (2000) examined soil chemical and biological characteristics across the vegetation and elevation gradient within the Coweeta basin. They found that vegetation was the dominant regulating factor of soil biological indices, such as soil respiration and N transformations. In the northeastern United States, Mitchell (2011) examined the relationship between site N export and other site factors. He found that while N deposition played a role, soils and vegetation were also an important factor. Soil Ca availability was a determining factor in vegetation composition and the abundance of sugar maple was related to high rates of N mineralization and nitrification.

In this study, we found increased soil solution NO_3-N and Ca concentrations 7 years postharvest in WS 7 30-cm tension lysimeters compared to WS 2 (figure 4.12). Ca concentrations in soil solutions collected at 100 cm were also significantly greater in WS 7 compared to WS 2 (figure 4.13). These data, along with stream chemistry data (see Swank et al., chapter 3, this volume), suggest that nutrient mobility increased following harvest. Nutrient movement is related to both the disturbance response and changes in water movement through the soil profile. As previously discussed, following the site harvest, we measured increased soil N availability as noted by greater NO_3-N and NH_4-N concentrations as well as potential N transformation rates, plus increased soil cation concentrations. Knoepp and Clinton (2009) examined changes in soil N availability and NO_3 in soil solution following site harvesting. They found increases in N only in harvested areas, and stream NO_3-N concentrations increases only in sites where all riparian zone vegetation had been removed.

The relationship between nutrient movement and water movement through forest ecosystems has been studied through the impacts of throughfall quantity and chemistry on soil solution and stream chemistry. Armbruster et al. (2002) examined Mg deposition patterns in throughfall and outputs in soil solution and streamflow in 71 forest ecosystems across Europe. They found a gradient of deposition, with the highest Mg inputs near the ocean but increased input up to 200 km inland. Soil solution and stream output fluxes were greater where inputs in throughfall were greatest with minor influence of site soils and bedrock. Several researches experimentally decreased the throughfall inputs to the forest floor (Johnson et al. 2002a; Neirynck et al. 2002) finding concomitant declines in soil water and nutrient fluxes. Johnson et al. (2002a) also experimentally increased throughfall amounts and found that while fluxes were greater than the artificially dry site, there was no significant difference from the ambient site.

Summary

We examined the effects of forest harvesting on nutrient inputs as slash and through-fall as well as responses of total soil C and N, extractable soil cations, nitrogen availability, and movement of nutrients out of the soil through leaching. Changes of nutrient inputs, due to the input of logging residue, and changes in throughfall resulted in increased total C and N and extractable cation pools in surface soils. These increases were not long-lived, although slash residue remained on site. On the other hand, soil extractable N and potential rates of N mineralization and nitrification remained elevated for more than 20 years after harvest. We concluded that changes in vegetation following harvesting have resulted in increased rates of N cycling. This continued rate of N cycling is evident not only in the soil N processes but also in continued stream export of NO_3-N (see Swank et al., chapter 3, this volume).

Literature Cited

Abbott, D. T., and D. A. Crossley, Jr. 1982. Woody litter decomposition following clear-cutting. *Ecology* 63: 35–42.
Alban, D. H. 1982. Effects of nutrient accumulation by aspen, spruce, and pine on soil properties. *Soil Science Society of America Journal* 46: 853–861.
Andersson, T. 1991. Influence of stemflow and throughfall from common oak (*Quercus robur*) on soil chemistry and vegetation patterns. *Canadian Journal of Forest Research* 21: 917–924.
Armbruster, M., J. MacDonald, N. B. Dise, and E. Matzner. 2002. Throughfall and output fluxes of Mg in European forest ecosystems: a regional assessment. *Forest Ecology and Management* 164: 137–147.
Baker, T. G., G. M. Will, and G. R. Oliver. 1989. Nutrient release from silvicultural slash-leaching and decomposition of *Pinus radiata* needles. *Forest Ecology and Management* 27: 53–60.
Billett, M. F., F. Parker-Jervis, E. A. Fitzpatrick, and M. S. Cresser. 1990. Forest soil chemical changes between 1949/50 and 1987. *Journal of Soil Science* 41: 133–145.
Binkley, D., D. Valentine, C. Wells, and U. Valentine. 1989. An empirical analysis of the factors contributing to 20-year decrease in soil pH in an old-field plantation of loblolly pine. *Biogeochemistry* 8: 39–54.
Boring, L. R., W. T. Swank, and C. D. Monk. 1988a. Dynamics of early successional forest structure and processes in the Coweeta basin. Pages 161–179 in *Forest Hydrology and Ecology at Coweeta*. W. T. Swank and D. A. Crossley, Jr., editors. Springer-Verlag, New York, New York.
Boring, L. R., W. T. Swank, J. B. Waide, and G. S. Henderson. 1988b. Sources, fates, and impacts of nitrogen inputs to terrestrial ecosystems: review and synthesis. *Biogeochemistry* 6: 119–159.
Bremner, J. M., and C. S. Mulvaney. 1982. Nitrogen-Total. Pages 595–624 in *Methods of soil analysis Part 2. Chemical and Microbiological Properties*. A. L. Page, R. H. Miller, and D. R. Keeney, editors. American Society of Agronomy and Soil Science Society of America, Madison, Wisconsin.

Cole, D. 1995. Soil nutrient supply in natural and managed forests. *Plant and Soil* 168–169: 43–53.

Dahlgren, R. A., and C. T. Driscoll. 1994. The effects of whole-tree clear-cutting on soil processes at the Hubbard Brook Experimental Forest, New Hampshire, USA. *Plant and Soil* 158: 239–262.

Donaldson, J., and G. Henderson. 1990. Nitrification potential of secondary-successional upland oak forests: I. mineralization and nitrification during laboratory incubations. *Soil Science Society of America Journal* 54: 892–897.

Eaton, J. S., G. E. Likens, and F. H. Bormann. 1973. Throughfall and stemflow chemistry in a northern hardwood forest. *Journal Ecology* 61: 495–508.

Edwards, N. T., and B. M. Ross-Todd. 1983. Soil carbon dynamics in a mixed deciduous forest following clear-cutting with and without residue removal. *Soil Science Society of America Journal* 47: 1014–1021.

Elliott, K. J., L. R. Boring, W. T. Swank, and B. L. Haines. 1997. Successional changes in plant species diversity and composition after clearcutting a Southern Appalachian watershed. *Forest Ecology and Management* 92: 67–85.

Fahey, T. J., J. W. Hughes, M. Pu, and M. A. Arthur. 1988. Root decomposition and nutrient flux following whole-tree harvest of northern hardwood forest. *Forest Science* 34: 744–768.

Federer, C. A., J. W. Hornbeck, L. M. Tritton, C. W. Martin, R. S. Pierce, and C. T. Smith. 1989. Long-term depletion of calcium and other nutrients in eastern US forests. *Environmental Management* 13: 593–601.

Gallardo, J. F., A. Martin, G. Moreno, and I. Santa Regina. 1998. Nutrient cycling in deciduous forest ecosystems of the Sierra de Gata mountains: nutrient supplies to the soil through both litter and throughfall. *Annales Des Sciences Forestieres* 55: 771–784.

Grover, B. L., and R. E. Lamborn. 1970. Preparation of porous ceramic cups to be used for extraction of soil water having low solute concentrations. *Soil Science Society of America Proceedings* 34: 706–708.

Hendrickson, O. Q. 1988. Biomass and nutrients in regenerating woody vegetation following whole-tree and conventional harvest in a northern mixed forest. *Canadian Journal of Forest Research* 18: 1427–1436.

Hendrickson, O. Q., L. Chatarpaul, and D. Burgess. 1989. Nutrient cycling following whole-tree and conventional harvest in northern mixed forest. *Canadian Journal of Forest Research* 19: 725–735.

Hursh, C. R. 1928. Litter keeps forest soil productive. *Southern Lumberman* 133: 219–221.

Johnson, D. W., P. J. Hanson, and D. E. Todd. 2002a. The effects of throughfall manipulation on soil leaching in a deciduous forest. *Journal of Environmental Quality* 31: 204–216.

Johnson, D. W., G. S. Henderson, and D. E. Todd. 1988. Changes in nutrient distribution in forests and soils of Walker Branch Watershed over an eleven-year period. *Biogeochemistry* 5: 275–293.

Johnson, D. W., J. D. Knoepp, W. T. Swank, J. Shan, L. A. Morris, D. H. Van Lear, and P. R. Kapeluck. 2002b. Effects of forest management on soil carbon: results of some long-term resampling studies. *Environmental Pollution* 116: S201–S208.

Johnson, D. W., D. E. Todd, C. F. Trettin, and J. S. Sedinger. 2007. Soil carbon and nitrogen changes in forests of walker branch watershed, 1972 to 2004. *Soil Science Society of America Journal* 71: 1639–1646.

Jones, J. B., Jr. 1977. Elemental analysis of soil extracts and plant tissue ash by plasma emission spectroscopy. *Communications in Soil Science and Plant Analysis* 8: 349–365.

Knoepp, J. D., and B. D. Clinton. 2009. Riparian zones in southern Appalachian headwater catchments: Carbon and nitrogen responses to forest cutting. *Forest Ecology and Management* 258: 2282–2293.

Knoepp, J. D., D. C. Coleman, D. A. Crossley, Jr., and J. S. Clark. 2000. Biological indices of soil quality: an ecosystem case study of their use. *Forest Ecology and Management* 138: 357–368.

Knoepp, J. D., and W. T. Swank. 1994. Long-term soil chemistry changes in aggrading forest ecosystems. *Soil Science Society of America Journal* 58: 325–331.

Knoepp, J. D., and W. T. Swank. 1997. Forest management effects on surface soil carbon and nitrogen. *Soil Science Society of America Journal* 61: 928–935.

Knoepp, J. D., J. M. Vose, B. D. Clinton, and M. D. Hunter. 2011. Hemlock infestation and mortality: impacts on nutrient pools and cycling in Appalachian forests. *Soil Science Society of America Journal* 75: 1935–1945.

Knoepp, J. D., J. M. Vose, and W. T. Swank. 2008. Nitrogen deposition and cycling across an elevation and vegetation gradient in southern Appalachian forests. *International Journal of Environmental Studies* 65: 389–408.

Kraske, C. R., and I. J. Fernandez. 1993. Biogeochemical responses of a forested watershed to both clearcut harvesting and papermill sludge application. *Journal of Environmental Quality* 22: 776–786.

Lapointe, B., R. L. Bradley, and B. Shipley. 2005. Mineral nitrogen and microbial dynamics in the forest floor of clearcut or partially harvested successional boreal forest stands. *Plant and Soil* 271: 27–37.

Lindberg, S. E., and J. G. Owens. 1993. Throughfall studies of deposition to forest edges and gaps in montane ecosystems. *Biogeochemistry* 19: 173–194.

Mann, L. K., D. W. Johnson, D. C. West, D. W. Cole, J. W. Hornbeck, C. W. Martin, H. Riekerk, C. T. Smith, W. T. Swank, L. M. Tritton, and D. H. Van Lear. 1988. Effects of whole-tree and stem-only clearcutting on postharvest hydrologic losses, nutrient captial and regrowth. *Forest Science* 34: 412–428.

Mattson, K. G., and W. T. Swank. 1989. Soil and detrital carbon dynamics following forest cutting in the southern Appalachians. *Biology and Fertility of Soils* 7: 247–253.

Mattson, K. G., W. T. Swank, and J. B. Waide. 1987. Decomposition of woody debris in a regenerating, clearcut forest in the Southern Appalachians. *Canadian Journal of Forest Research* 17: 712–721.

Mitchell, M. J. 2011. Nitrate dynamics of forested watersheds: spatial and temporal patterns in North America, Europe and Japan. *Journal of Forest Research* 16: 333–340.

Mroz, G. D., M. F. Jurgensen, and D. J. Frederick. 1985. Soil nutrient changes following whole tree harvesting on three northern hardwood sites. *Soil Science Society of America Journal* 49: 1552–1557.

Mueller, C. W., and I. Kogel-Knabner. 2009. Soil organic carbon stocks, distribution, and composition affected by historic land use changes on adjacent sites. *Biology and Fertility of Soils* 45: 347–359.

Neirynck, J., E. Van Ranst, P. Roskams, and N. Lust. 2002. Impact of decreasing throughfall depositions on soil solution chemistry at coniferous monitoring sites in northern Belgium. *Forest Ecology and Management* 160: 127–142.

Nelson, D. W., and L. E. Sommers. 1982. Total carbon, organic carbon, and organic matter. Pages 539–580 in *Methods of Soil Analysis Part 2. Chemical and Microbiological Properties*. A. L. Page, R. H. Miller, and D. R. Keeney, editors. American Society of Agronomy and Soil Science Society of America, Madison, Wisconsin.

O'Connell, A. M., T. S. Grove, D. S. Mendham, and S. J. Rance. 2004. Impact of harvest residue management on soil nitrogen dynamics in *Eucalyptus globulus* plantations in south western Australia. *Soil Biology and Biochemistry* 36: 39–48.

Paul, K. I., P. J. Polglase, and G. P. Richard. 2003. Predicted change in soil carbon following afforestation or reforestation, and analysis of controlling factors by linking a C accounting model (CAMFor) to models of forest growth (3PG), litter decomposition (GENDEC) and soil C turnover (RothC). *Forest Ecology and Management* 177: 485–501.

Potter, C. S. 1992. Stemflow nutrient inputs to soil in a successional hardwood forest. *Plant and Soil* 140: 249–254.

Potter, C. S., H. L. Ragsdate, and W. T. Swank. 1991. Atmospheric deposition and foliar leaching in a regenerating southern Appalachian forest canopy. *Journal of Ecology* 79: 97–115.

Richter, D. D., D. Markewitz, C. G. Wells, H. L. Allen, R. April, P. R. Heine, and B. Urrego. 1994. Soil chemical change during three decades in an old-field loblolly pine (*Pinus taeda* L.) ecosystem. *Ecology* 75: 1463–1473.

SAS. 2000. SAS version 9.1, SAS systems for Windows. SAS Institute Inc., Cary, North Carolina.

Scott, N. A., C. A. Rodrigues, H. Hughes, J. T. Lee, E. A. Davidson, D. B. Dail, and P. Malerba. 2004. Changes in carbon storage and net carbon exhange one year after an intital shelterwood harvest at Howland Forest, ME. *Environmental Management* 33 (Supplement): S9–S22.

Shammas, K., A. M. O'Connell, T. S. Grove, R. McMurtrie, P. Damon, and S. J. Rance. 2003. Contribution of decomposing harvest residues to nutrient cycling in a second rotation *Eucalyptus globulus* plantation in south-western Australia. *Biology and Fertility of Soils* 38: 228–235.

Smethurst, P. J., and E. K. S. Nambiar. 1990. Distribution of carbon and nutrients and fluxes of mineral nitrogen after clear-felling a *Pinus radiata* plantation. *Canadian Journal of Forest Research* 20: 1490–1497.

Snyder, K. E., and R. D. Harter. 1984. Changes in solum chemistry following clearcutting of northern hardwood stands. *Soil Science Society of America Journal* 48: 223–228.

Swank, W. T., and G. S. Henderson. 1976. Atmospheric input of some cations and anions to forest ecosystems in North Carolina and Tennessee. *Water Resources Research* 12: 541–546.

Swank, W. T., L. J. Reynolds, and J. M. Vose. 1992. Annual mean atmospheric deposition, throughfall, stemflow, and soil solution fluxes for major ions at Coweeta Hardwood Site. Page 634 in *Atmospheric Deposition and Forest Nutrient Cycling*. D. W. Johnson and S. E. Lindberg, editors. Springer-Verlag, New York, New York.

Swank, W. T., J. Swift, L.W., and J. E. Douglass. 1988. Streamflow changes associated with forest cutting, species conversions, and natural disturbances. Pages 297–312 in *Forest Hydrology and Ecology at Coweeta*. W. T. Swank and D. A. Crossley, Jr., editors. Springer-Verlag, New York, New York.

Swank, W. T., and J. M. Vose. 1988. Effects of cutting practices on microenvironment in relation to hardwood regeneration. Pages 71–88 in *Guidelines for Regenerating Appalachian Hardwood Stands: proceedings of a workshop*. 24-26 May 1988, Morgantown, WV. SAF Publication 88-034. West Virginia University Books, Office of Publications, Morgantown.

Thomas, D. J. 1996. Soil Survey of Macon County, North Carolina. Soil Survey, USDA Natural Resource Conservation Service.

Tiedemann, A. R., T. M. Quigley, and T. D. Anderson. 1988. Effects of timber harvest on stream chemistry and dissolved nutrient losses in northeast Oregon. *Forest Science* 34: 344–358.

Waide, J. B., W. H. Caskey, R. L. Todd, and L. R. Boring. 1988. Changes in soil nitrogen pools and transformations following forest clearcutting. Pages 221–232 in *Forest Hydrology and Ecology at Coweeta*. W. T. Swank and D. A. Crossley, Jr., editors. Springer-Verlag, New York, New York.

5

Soluble Organic Nutrient Fluxes

Robert G. Qualls*
Bruce L. Haines
Wayne T. Swank

Introduction

Inorganic nutrients have been the focus of most studies of the cycling and leaching of elements after disturbance. However, soluble organic nutrients, such as the forms of C, N, and P that are bound in organic matter, are also released from living and dead organic matter. The mechanisms by which inorganic nutrients are retained or lost after clearcutting are generally well known and illustrated in many studies. These include loss of root uptake (Likens and Bormann 1995), rapid recovery of root uptake by stump sprouts (Boring et al. 1988), recovery of root uptake by seedling growth (Marks 1974), delayed mineralization and subsequent nitrification due to a high C/N ratio in litter (Vitousek et al. 1979), temporary sorption on ion exchange sites (Vitousek et al. 1979), and in the case of P, fixation or sorption on soil (Wood et al. 1984; Walbridge et al. 1991). The increase in water flux from the root zone due to cutting and the concomitant reduction in evapotranspiration also plays an important role in controlling the leaching of nutrients (Likens and Bormann 1995). The factors that control the leaching of *organic* nutrients after clearcutting or other disturbances, however, have not been extensively investigated.

Dissolved organic nitrogen (DON) is the major form of N in the streamwater draining many mature forest watersheds (Lewis 2002; Perakis and Hedin 2007). Relatively high concentrations of DON drain from the forest floor, and this DON also generally makes up most of the total N draining from the forest floor of intact forests (Qualls et al. 1991; Michalzik et al. 2001). The importance of DON in solution transport in intact forests and the sudden inputs of potentially soluble nutrients in logging slash suggest that the transport of soluble organic nutrients may be important in the retention or loss of nutrients after clearcutting.

Our objectives in this study were to (i) compare fluxes of the dissolved organic nutrients dissolved organic carbon (DOC), DON, and dissolved organic phosphorus

* Corresponding author: Department of Natural Resources and Environmental Science, University of Nevada, M.S. 370, Reno, NV 89557 USA

(DOP) in a clearcut area and an adjacent mature reference area, (ii) determine whether concentrations of dissolved organic nutrients or inorganic nutrients were greater in clearcut areas than in reference areas, and (iii) identify the strata where the greatest net leaching and deposition occur.

Site and Methods

The study site was on, or adjacent to, Watershed 2 (WS 2) at the Coweeta Hydrologic Laboratory in the Nantahala Range of the Southern Appalachian Mountains of North Carolina (83°26'W, 35°04'N) at an elevation of 840 m. Annual precipitation was 127.6 and 153.4 cm during the first and second years of the study, respectively. Snow comprises only 2% to 10% of precipitation.

The area was covered by a deciduous forest dominated by several species of *Quercus*, *Carya* spp., *Acer rubrum*, and *Cornus florida*. The forest had been undisturbed for at least 62 years except for mortality due to the chestnut blight (Monk and Day 1988). Thickets of *Kalmia latifolia* and *Rhododendron maximum* cover portions of the study area. Soil in the study area is Chandler loam, a coarse-loamy, micaceous, mesic, Typic Dystrochrept. The dry mass of the forest floor on WS 2 averaged 1145 g/m² (Ragsdale and Berish 1988). Annual litterfall was 498 g/m² (dry mass) and had a C/N mass ratio of 60 (W. T. Swank, unpublished data, 1991).

An experimental clearcutting was combined with the installation of a weather station in an area on the perimeter of WS 2 to simulate the clearcutting experiment on the adjacent WS 7. An area of 890 m² was cut in November 1985, after leaf fall. Four 5 m x 5 m plots were randomly located within the area, excluding the weather station. The perimeter of the clearcut area was trenched to about a 60 cm depth, and the trench was lined with plastic to prevent root growth from the surrounding forest. We uniformly redistributed woody debris over the plots so that the dry-weight equivalent of approximately 120 Mg/ha lay on each plot to mimic the experimental clearcut on WS 7 in 1977 (Boring et al. 1988). Then, an uncut reference plot was randomly located in the area, or areas, matching all criteria for slope, aspect, soil series, and depth of the A horizon for a given cut plot. Thus, each cut plot was paired with an uncut plot and treated as a block, as in a case-control experimental design (Breslow 1996).

Solution was collected above the forest floor (throughfall or slash leachate), below the Oa horizon, in the mid-A horizon, the mid-AB horizon, the mid-B, and 20 cm below the upper boundary of the C horizon. In the cut plots, slash leachate collectors were placed *above* the litter but beneath all woody logging debris. Sampling and the measurement of water fluxes were described by Qualls et al. (2000). Water fluxes in throughfall and from the bottom of the Oa horizon were measured as by Qualls et al. (1991). Interception by forest floor litter in the clearcut was assumed to be the same as in the reference plots. Annual water fluxes from the bottom of the rooting zone of the uncut plots were assumed to be equal to the annual streamflow on the gauged watershed (WS 2). Using this flux as a reference, annual fluxes from the rooting zone of the cut plots were based on an empirical model that predicts the increase in streamflow due to cutting over that of a reference watershed at Coweeta (Douglass and Swank 1975). Water fluxes from each depth increment between the bottom of the Oa horizon and the bottom of the root zone were interpolated by distributing total transpiration among soil increments in proportion to the distribution of fine roots (McGinty 1976). Fluxes for each form of nutrient were then calculated by multiplying the water flux by the concentration of the each nutrient form.

Concurrent with the clearcutting study, a larger study involving an additional eight plots on the reference WS 2 was done that also measured fluxes in streamwater and a more detailed examination of mechanisms. Other aspects of this study that have been presented include: annual fluxes of C, N, and P from throughfall and from the forest floor of the uncut area of WS 2 (Qualls et al. 1991), potential rates of biodegradation of DOC and DON from all strata (Qualls and Haines 1992b), chemical fractionation of DOC and DON from all strata (Qualls and Haines 1991), measurement of adsorption of DOC (Qualls and Haines 1992a), determination of the mechanisms of adsorption of DOC (Qualls 2000), effects of clearcutting on DOC, DON, and DOP concentrations (Qualls et al. 2000), and an analysis of the factors controlling fluxes through the soil and from streamwater on the uncut watershed (Qualls et al. 2002).

Results and Discussion

Over a two-year period the estimated water flux from the bottom of the rooting zone was 1.47 times higher in the cut plots, or 26 cm (table 5.1). This increase in water flux is very similar to streamflow increases measured on adjacent WS7 (23 cm yr^{-1}) the first two years following clearcutting (Swank et al. 2001; see also Swank et al., chapter 3, this volume).

Concentrations and fluxes of DOC, N forms, and P forms in the cut versus uncut plots (figures 5.1–3) demonstrated three major points: First, dissolved organic C and N concentrations were higher in the cut plots in slash leachate (vs. throughfall), forest floor leachate, A horizon soil solution, and B horizon soil solution (figures 5.1–2). In the case of DOP, concentrations were much higher in the cut

Table 5.1 Average hydrologic fluxes over the two-year sampling period (in annual units).

Stratum or horizon	Water fluxes (m/yr)	
	Uncut	Cut
Precipitation	1.42	1.42
Throughfall or slash	1.25	1.36
Oa	1.22	1.32
A	0.89	1.07
AB	0.74	0.95
B	0.58	0.83
C	0.55	0.81

Source: Qualls et al. (2000).

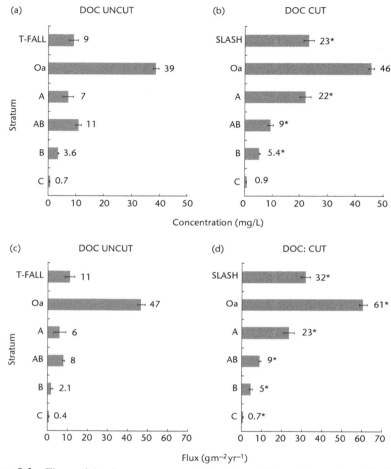

Figure 5.1   Flux weighted average concentrations (**a** and **b**) and fluxes (**c** and **d**) of DOC for the uncut and cut plots. Strata or soil horizons are indicated on the vertical axis. Asterisks indicate significant ($P < 0.05$) differences (a significant main effect of cutting treatment in the ANOVA) between cut vs. uncut plots and are placed on the bar that was greater in magnitude. Error bars are standard error of the mean, indicated only for the organic form, and reflect variability among plots, not temporal variability. Numbers beside the bars indicate values. From Qualls et al. (2000).

plots in the slash leachate (vs. throughfall) and forest floor but not in the mineral soil (figure 5.3). Second, greater water fluxes through the soil horizons of the cut plots (table 5.1) combined with greater concentrations in some horizons to give greater fluxes of DOC, DON, and DOP in all strata (figures 5.1–3). Third, fluxes of DON were greater than those of dissolved inorganic N, even in the cut plots (figure 5.2). However in the case of P, fluxes of inorganic P exceeded those of DOP in the cut plots in slash and forest floor leachate (figure 5.3).

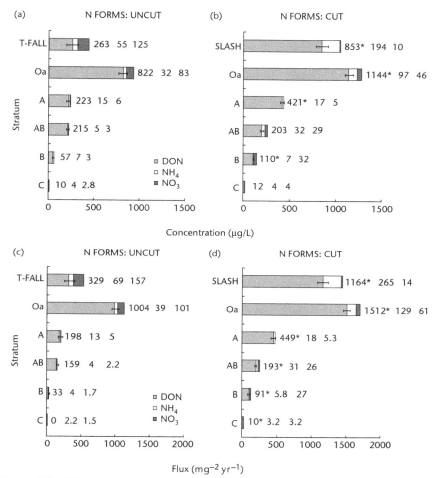

Figure 5.2 Flux weighted average concentrations (figures **a** and **b**) and fluxes (figures **c** and **d**) of N forms for the uncut and cut plots. Asterisks indicate significant ($P < 0.05$) differences (a significant main effect of cutting treatment in the ANOVA) in DON (not inorganic forms) between cut vs. uncut plots, and are placed on the bar that was greater in magnitude. Error bars are standard error of the mean, indicated only for the organic form, and reflect variability among plots, not temporal variability. Numbers on or beside the bars indicate values and are in the same order as the stacking of the bars. From Qualls et al. (2000).

Sources of DOM above the Mineral Soil

Sources of DOM in the cut plots were slash from cutting, other organic debris on the forest floor, and perhaps litter from dead roots. On the other hand, leaching from live canopy leaves and litterfall during the first two years after cutting was greatly reduced. In the mature forest canopy, leaching was an important source of DOC and, in particular, DOP (Qualls et al.1991). In the cut plots, however, fluxes of DOC, DON, and DOP in slash throughfall were much higher than in

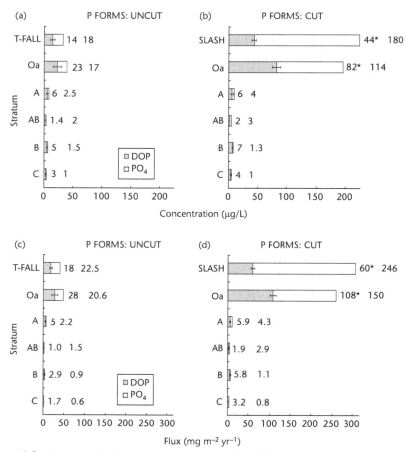

Figure 5.3   Flux weighted average concentrations (**a** and **b**) and fluxes (**c** and **d**) of P forms for the uncut and cut plots. Asterisks indicate significant ($P < 0.05$) differences (a significant main effect of cutting treatment in the ANOVA) in DOP (not PO_4) between cut vs. uncut plots, and are placed on the bar that was greater in magnitude. Error bars are standard error of the mean, indicated only for the organic form, and reflect variability among plots, not temporal variability. Numbers on or beside the bars indicate values and are in the same order as the stacking of the bars. From Qualls et al. (2000).

throughfall in the uncut plots. Sources of this throughfall in the cut plots may have included: (i) leaching of tannins from dead and fragmented bark, (ii) leaching of soluble organics from porous and fragmented wood, (iii) dissolution of lignin and other constituents by microbial enzymes, and (iv) leaching of microbial biomass such as that of shelf fungi.

Despite greater fluxes from the Oa horizon in the cut plots for all organic nutrients, there was less net leaching (defined as flux from the Oa horizon minus that in throughfall or slash) of DOC and DON from the forest floor in the cut plots compared to the uncut plots. Net leaching of DOC and DON from the Oa horizon was 19% and 48% lower, respectively, in the cut plots compared to the uncut plots (figures 5.1a and 5.2a vs. 5.1b and 5.2b). This was likely due to the loss of most new leaf litter production after cutting.

In a study following three years after clearcutting in a boreal forest, Piirainen et al. (2002) found that DOC and DON fluxes from the Oa horizon nearly doubled, results similar to those of our study but with somewhat greater increases. As in our study, fluxes of DON were greater than those of inorganic N in forest floor of the boreal clearcut. In a Norway spruce forest after clearcutting, Smolander et al. (2001) found that concentrations of DON percolating from the Oa horizon were only 17% higher in clearcut plots compared to intact forest, but concentrations were much higher at the 10 cm depth below the Oa horizon. In another clearcutting experiment in a *Picea abies* forest, Kalbitz and Bol (2004) removed all logging debris, thus removing canopy leaching, fresh litterfall, and logging debris as sources of DOM. Despite the elimination of logging debris in the cut plots, they still found slightly increased concentrations and increased fluxes of DOC and DON from the Oa horizon, mainly due to increased water flux. They attributed this effect to increases in temperature and a greater decomposition rate in the forest floor in cut plots. The Oe and Oa horizons were much thicker in the plots studied by Kalbitz and Bol than in the Coweeta study, and that could have contributed to a more sustained source of DOC than in the Coweeta study. Dai et al. (2001) examined a forest that had been clearcut 15 years earlier at the Hubbard Brook Experimental Forest and found that concentrations of DOC from the forest floor were still much higher than in the intact forest and that the DOC being leached was more aromatic in chemical nature. This chemical difference might reflect the sustained leaching of woody debris and the reduction in canopy leaching that contains more labile, but less aromatic DOC (Qualls et al. 1992b). Mattson et al. (1987) found that concentrations of DOC still averaged 76 mg/L in leachate from decaying logs seven years after clearcutting on the adjacent WS 7 at Coweeta. This suggests that decaying woody residue may remain a source of dissolved organic matter for several years.

The leaching of fine root litter after senescence can be a major source of DOC and DON (Uselman et al. 2007). The mortality of fine roots caused by clearcutting was not measured in this study, but the inputs to the A and AB horizons after clearcutting could contribute to higher concentrations observed in this and other studies (Smolander et al. 2001; Piirainen et al. 2002).

Removal of DOM in the Mineral Soil

Concentrations of DOC and DON declined with depth in the mineral soil, and the greatest difference between the cut and uncut plots occurred in the A horizon (figures 5.1 and 5.2). Physicochemical adsorption, largely by iron and aluminum oxyhydroxides, can rapidly remove DOC from solution and can buffer differences in input concentration (McDowell and Wood 1984; Qualls and Haines 1992b). It is unlikely that large proportions of the DOC and DON were removed by decomposition in the dissolved phase because DOC and DON from the uncut plots was very slow to mineralize (Qualls and Haines 1992b). The unusual degree of retention of soluble organic matter in WS 2 can be explained, in part, by the unusually high content of potentially adsorbing Fe and Al oxyhydroxides. The AB horizon soil used in the adsorption experiments by Qualls and Haines (1992a) had an oxalate extractable Fe and Al content of 1.8 and 2.7 g/kg, respectively, and a citrate bicarbonate dithionite extractable Fe and Al content of 22 and 12 g/kg, respectively. The AB horizon of this same soil from a nearby plot had the highest total Fe and Al contents of all 19 sites in the Integrated Forest Study (April and Newton 1992).

Few studies have measured fluxes of DOM below the forest floor, but Piirainen et al. (2002) found, as we did, that despite much higher fluxes of DOC and DON from the organic horizons these dissolved organic nutrients were mainly retained by the mineral soil in clearcut plots. In a clearcut and control *Pseudotsuga menziesii* forested watershed in Oregon, Sollins and McCorison (1981) monitored DOC (second and third year after cutting only) and DON (third year after cutting only) in soil solution. They found that concentrations of DOC were higher in soil solution in the clearcut by factors ranging from 1.4- to 1.9-fold. The DON comprised from 41% to 58% of total N in soil solution in the third year after cutting. Like the Coweeta site, this clearcut forest exhibited a lag in nitrification and nitrate concentrations that generally remained well below the 1 mg/L level.

In the case of DOP and PO_4, the relatively high concentrations draining from the forest floor of the cut plots were abruptly reduced to low levels in the A horizon, levels that were similar to those of the uncut plots (figure 5.3 a and b). This may reflect the strong tendency of PO_4 (Walbridge et al. 1991) and perhaps organic phosphate esters to adsorb in these Fe- and Al-rich soils.

The increase in water fluxes through the soil was an important factor in causing greater fluxes of organic nutrients from the lower soil horizon; these were more important, in fact, than differences in concentration. This close relationship between nutrient output and water flux is well known for inorganic ions at the watershed level, such as Ca and Na, where concentration is relatively constant and flux is proportional to streamflow (Swank 1988; Likens and Bormann 1995). The estimated increases in annual water flux of 26 cm (a factor of 1.47) due to the cutting of our plots (table 5.1) lies within the ranges found in several studies (Sollins and McCorison 1981; Swank et al. 1988; Likens and Bormann 1995; Arthur et al. 1998).

Streamwater Fluxes of DOC, DON, and DOP

Concurrent with the study of the clearcut plots, the fluxes (export) of DOC, DON, and DOP were measured on the reference watershed (WS 2). Concentration of each species versus streamflow was modeled to estimate the fluxes over the two-year study period. In the case of DOC, during baseflow, DOC concentrations were relatively consistent, averaging 0.63 ($\pm$ 0.1 s.d.) mg/L and showed no seasonal trends. As a first approximation, a simple model of DOC concentration versus stormflow was able to accurately fit the DOC data. The water was assumed to be a mixture of water from two sources: baseflow with a constant concentration of 0.63 mg/L and stormflow (superimposed on baseflow during storms) with a DOC concentration of 5.0 mg/L when rising and peaking and 3.9 mg/L when falling (based on regressions). Plotting DOC concentration versus the ratio of stormflow/baseflow yielded a linear regression line with a y intercept corresponding approximately to the concentration in baseflow (~0.6 to 0.8 mg/L) and the concentration in stormflow corresponding to y at x = 1 (100% stormflow). Fits to this simple model with a simple interpretation were very good, (r^2 = 0.83 for rising and 0.77 for falling limbs) but a slightly curvilinear relationship provided a better fit than the linear relationship. Patterns for DON were similar but more variable since DON concentrations were closer to the limit of detection in streamwater.

Fluxes of DOC, DON, and DOP, NH_4, NO_3, and PO_4 in streamwater at the weir of WS 2 are shown in table 5.2. DON comprised 79% of the total dissolved N in streamwater at the weir and 40% of the total N. DOP comprised about 46% of the total dissolved P. Tate and Meyer (1983) showed that four watersheds at the Coweeta Hydrologic Laboratory had a lower export of dissolved organic carbon (DOC) per unit runoff of water than all (15) other watersheds in studies reviewed. In addition, Meyer and Tate (1983) and Meyer et al. (in chapter 6 of this volume) found that the DOC export in the third and fourth year after clearcutting of WS 7 was somewhat lower than the control stream, perhaps due to reduced litter inputs to the stream and near stream source areas. Likewise, the export of DON from WS 2 in our study was unusually low. While the flux of DON from the C horizon of the cut plots was nearly double that of the uncut plots in our study, the contribution of these fluxes from the C horizon would be small in comparison to the DON export of many intact forest watersheds. The mean export of DON and total N from 19 minimally disturbed watersheds in the USA was 1.24 and 2.62 kg ha^{-1} yr^{-1} respectively (Lewis 2002), about 6.5 times that of WS 2. Those studies included some watersheds with considerable wetland area. The mean export of DON from 20 undisturbed tropical watersheds was 2.40 kg ha^{-1} yr^{-1} (Lewis et al. 1999). In the tropical watersheds DON comprised an average of 67% of total dissolved N in first- and second-order streams but was about 50% for all watersheds.

Table 5.2   Fluxes of dissolved organic and inorganic nutrients from the reference watershed (WS 2).

	Flux (kg ha^{-1} yr^{-1})
DOC	4.1
Particulate organic C*	3.6
Total organic C	7.7
DON	0.19
NO_3-N	0.036
NH_4-N	0.014
Particulate N**	0.23
Total N	0.47
DOP	0.011
PO_4-P	0.013

Note: All units are kg ha^{-1} yr^{-1}, unlike the figures.
* Estimated using fluxes from Swank and Waide (1988)
** Estimated using fluxes from Monk (1975)
Source: From data reported by Qualls et al. (2002)

Particulate N comprised only 17% of total N in first- and second-order streams ranging to 37% in rivers of the highest order. In a study of nine forested water-sheds in New England, the export of DON ranged from 0.5 to 2.4 kg ha^{-1} yr^{-1} with DON comprising the majority most of the total dissolved nitrogen (Campbell et al. 1999). The reasons for this watershed to be unusually retentive for dissolved organic nutrients lay in the high adsorption capacity of the soil and the tendency for most water to drain through the B horizon before entering the stream, as dis-cussed in the following sections.

Retention of Dissolved Organic Nutrients as a Function of Soil Type and Hydrologic Flowpath

The hypothetical relationship of hydrologic flowpath and soil adsorption capacity to the tendency of dissolved organic nutrients to leach from the ecosystem can be illustrated graphically. The diagram in figure 5.4 depicts geochemical and hydro-logical controls that dominate the tendency of an ecosystem to retain soluble organic nutrients produced by biological processes. Ecosystems can be compared on this diagram with respect to these characteristics. Geochemical processes controlling retention are largely dependent on the presence or absence of Fe and Al oxyhy-droxides or certain clays. One end member of this series along the geochemical axis might be represented by sand dunes and other sandy soils, such as the Indiana Dunes chronosequence examined by (Olson 1958). Another end member might be represented by soils high in oxyhydroxides (such as at Coweeta) or volcanic soils with allophane that strongly adsorb humic substances. Hydrologic bypassing cir-cuiting of B horizons high in metal oxyhydroxides can also bypass the adsorbing effects of soils, represented in the extreme by surface flow or surface flow wetlands. Streams may even be visualized within this framework, as a case of surface flow. The potential decrease in soluble-organic-matter production after cutting is another factor determining export and might be represented along an axis perpendicular to the other two axes in figure 5.4.

Comparison of Mechanisms Controlling Leaching of Dissolved Organic and Inorganic Nutrients

Numerous studies have demonstrated that leaching of inorganic N or P is greater in recently clearcut forests compared to mature reference stands (Sollins and McCorison 1981; Adamson et al. 1987; Stevens and Hornung 1990; Likens and Bormann 1995; Ring 1995). In this study, we also found that fluxes of dissolved organic nutrients were greater in clearcut plots. Indeed fluxes of NO$_3$, NH$_4$, and PO$_4$ were elevated in our cut plots, but the average concentrations did not approach the levels found for NO$_3$ in, for example, some cut forests (Likens and Bormann 1995). Partly because of this relatively small increase in NO$_3$ concentrations, the fluxes of DON typically remained greater than those of inorganic forms. The adjacent water-shed (WS 7) was experimentally clearcut in 1977 and NO$_3$ export in streamwater during the first and second year was only about 0.3 and 1.1 kg/ha, respectively (Swank 1988). In our cut plots, the flux of NO$_3$ from the C horizon was much lower

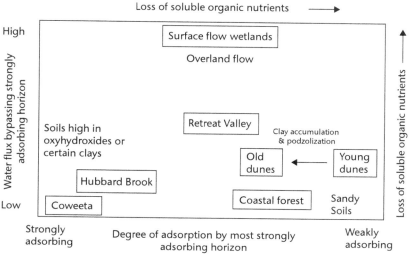

Figure 5.4 Classification of the ecosystems in their tendency to retain soluble organic nutrients as a function of the degree of adsorption of mineral soil and hydrological short circuiting. Ecosystems are placed on this diagram in relative positions since the data needed to quantify their position on the axes were generally not available. "Coweeta" refers to the current study, "Hubbard Brook" to the studies by McDowell and Wood (1984) and McDowell and Likens (1988) who suggested a somewhat lower degree of adsorption of DOC, "Coastal forest" to the study by Seely et al. (1998) where sandy texture appeared to provide a relatively low degree of adsorption but where hydrologic by-passing was not suggested, "Retreat Valley" to the study of Nelson et al. (1996) in which sandy soils overlaying clay soils and portions of the watershed being "poorly drained" suggested some degree of hydrologic by-passing of the clay layer. The hypothetical example of sand dune soil development could be represented by the Indiana Dunes chronosequence (Olson 1958). The "surface flow wetlands" or "overland flow" entry on the diagram represents the extreme example of hydrologic short circuiting which can bypass the adsorbing effects of soils (e.g., Qualls and Richardson 2003). From Qualls et al. (2002).

than that from the B horizon for unknown reasons, but the flux from the B horizon in our cut plots (0.27 kg/ha) was similar to export in streamwater during the first year after cutting on WS 7. However, the export from WS 7 in streamwater the second year after cutting was considerably higher than that from the B horizon in our cut plots. This relatively low export of inorganic nutrients was due to a very rapid recovery of root uptake in stump sprouts and herbaceous plants which recovered to 93% of the precutting N uptake in aboveground NPP only three years after cutting on mesic sites (Boring et al. 1988). A lag in nitrification may also have played a role in delaying nitrate loss from our cut plots, as in the studies of Vitousek et al. (1979). Output of N, especially NO_3, from clearcut forested watersheds varies by nearly two orders of magnitude (Vitousek et al. 1979; Emmett et al. 1990; Ring 1995). Although the data on leaching of dissolved organic nutrients after clearcutting are extremely limited, we hypothesize that the range of increase in concentrations and fluxes of DON is much less than that observed for NO_3.

A set of hypotheses comparing the factors controlling the retention of soluble organic versus inorganic nutrients (table 5.3; Qualls 2000) is applicable to our study. The nutrients considered are forms of nitrogen, phosphorus, and organic carbon only. In the case of the soluble organic nutrients, the generalizations are applied to macromolecules to exclude the free amino acids because they comprise a small percentage of the DON and because some plants can take up the smallest amino acids (Kielland 1994).

Perhaps the most important property of the inorganic N and P ions is their small molecular size, which allows transport through cell membranes. In contrast, the soluble macromolecules that carry most of the DOC, DON, and DOP do not pass through the cell membrane without being hydrolyzed first, which in turn requires extracellular decomposition for the assimilation of the nutrient element by microbes

Table 5.3   Factors controlling retention of soluble macromolecular organic vs. inorganic nutrients in terrestrial ecosystems: hypotheses.

Inorganic	Organic
Sources	
Microbial mineralization	Leaching from detritus
Atmospheric input	Direct leaching from plants, exudation
Direct leaching from plants	Microbial dissolution
Properties of molecules	
Small + and - ions	Mostly large molecules
Many salts soluble	Mostly—charged
Some salts insoluble (e.g. salts of PO_4)	Some molecules neutral
	Carboxyl group interactions important
	Multidentate bonding
	Most N in molecules does not act as cation
Removal from solution: Biological	
Root uptake	
Microbial uptake (immobilization)	Microbial hydrolysis and uptake of small molecules
Removal from solution: Nonbiological	
+ Electrostatic	Ligand exchange (regulating concentrations at a low level in mineral soil)
Ligand exchange ($H_2PO_4^-$)	H-Bonding or van der Waals forces
– Electrostatic (minor)	(regulating concentrations at a high level in organic horizons)
Chemical precipitation	
Major factors allowing loss from ecosystem	
Hydrologic short circuiting of root network or adsorbing soil horizon	Hydrologic short circuiting of adsorbing soil horizon
Removal of root uptake	Root uptake less important than for inorganic molecules, only small molecules.
Weak geochemical sorption/ precipitation potential of soil	Absence of a horizon high in Fe and Al oxyhydroxides and certain clays

Source: Qualls (2000).

and roots. Consequently, root uptake and direct microbial uptake, which are important in preventing the loss of soluble inorganic nutrients, are not factors for the macromolecular dissolved organic nutrients. Hence, geochemical factors are more important in controlling the leaching of dissolved organic nutrients.

Electrostatic charge is another property of the predominant soluble inorganic forms of N and P, making them susceptible to sorption on cation or anion exchange sites. Many of the salts formed with counter ions are soluble, but some, such as the calcium salts of P at high pH are insoluble. In addition, the presence of hydroxyl group on the phosphate ions make them susceptible to ligand exchange, which often may be the most important factor in preventing the leaching of phosphate ions.

Properties of the soluble organic macromolecules besides size that determine their behavior are (i) that they are predominately negatively charged, although a significant fraction is neutral (Qualls and Haines 1991); (ii) that the presence of carboxyl and phenolic hydroxyl groups make such interactions as ligand exchange and hydrogen bonding important; and (iii) that molecules are multidentate, making bonds more stable. In addition, the N atoms in the humic and hydrophilic acids do not contribute substantial positive charges in the macromolecules, as they do in peptides. Instead, the carboxyl and phenolic hydroxyl groups largely determine the behavior of the N carried more or less "passively" by the humic and hydrophilic acids (Qualls and Haines 1991). In the case of dissolved organic P, most macromolecules containing P behave as anions, but whether the negatively charged P ester groups or the carboxylic acids determine this behavior has not been determined (Qualls and Haines 1991).

As in the case for phosphate, ligand exchange is likely to be responsible for the removal of a large portion of the macromolecular dissolved organic molecules in mineral soils (Qualls 2000). Thus the geochemical mechanisms for retaining phosphate, DOC, DON, and DOP are similar. These mechanisms are capable of maintaining relatively low levels in solution. Organic-organic mechanisms, such as hydrogen bonding or van der Waals forces, may also remove these macromolecules in organic horizons, but these mechanisms function to maintain concentrations at higher levels (Qualls 2000).

We can classify the various mechanisms of retention as *geochemical, hydrologic,* and *biological.* In the case of N, the mechanisms controlling the loss of N in the form of nitrate are largely biological and hydrologic. We propose that the loss of DON is controlled by geochemical and hydrologic mechanisms. The production of soluble organic nutrients is, of course, biological, but dissolution and sorption are geochemical mechanisms.

We hypothesize that the most important geochemical mechanisms leading to the retention of dissolved organic nutrients are (i) the slow, sustained release of potentially soluble organic matter caused by slow dissolution, equilibrium-controlled desorption from organic surfaces, and the gradual exposure of surfaces to percolating water during fragmentation; and (ii) equilibrium adsorption to Fe and Al oxyhydroxides and clays. The slow, gradual release of potentially soluble organic matter from detritus can be compared to factors tending to delay nitrification, as in the studies of Vitousek et al. (1979). Sorption helps retain the soluble organic matter to be decomposed slowly on surfaces and finally, hydrologic factors control the capacity for this adsorption capacity to be effective at retaining these organically bound forms of nutrients.

Conclusions

Concentrations of DOC and DON were higher in the cut plots than in uncut plots in solutions from slash leachate (vs. throughfall), the forest floor, the A horizon, and the B horizon. DOP concentrations were higher in the cut plots than in the uncut plots in solutions from slash leachate (vs. throughfall) and the forest floor but not in the mineral soil.

Fluxes of DOC, DON, and DOP in all strata were greater in cut plots than in uncut plots, a product not only of concentration differences in some cases, but also a 1.47-fold greater flux of water. Even in the cut plots, fluxes of the organic forms of nutrients exceeded those of the inorganic forms (except in the case of P in slash leachate and forest floor solution).

Despite greater fluxes of dissolved organic N from the cut plots, over 99% of the DON draining from the forest floor on the cut plots was removed (presumably adsorbed) above the upper C horizon, demonstrating a remarkable degree of retention of this soluble form of N. We hypothesize that the well-recognized retention mechanisms for inorganic nutrients (e.g., uptake by the roots of stump sprouts, adsorption of ions, and immobilization) combined with geochemical adsorption of dissolved organic matter, efficiently buffer against the leaching of either soluble inorganic or organic nutrients after clearcutting.

Acknowledgments

We thank James Vose for supplying meteorological data for PROSPER, Larry Morris for use of his soil physics lab, and Julia Gaskin for advice. We also thank Kent Tankersley, Steve Wooten, and Lisa Leatherman for field and lab assistance. George Fernandez of the University of Nevada provided valuable statistical advice. Our work was supported by NSF Grants BSF-8501424 and BSF-8514328.

Literature Cited

Adamson, J. K., M. Hornung, D. G. Pyatt, and A. R. Anderson. 1987. Changes in solute chemistry of drainage waters following clearfelling of a Sitka spruce plantation. *Forestry* 60: 165–177.

April, R., and R. Newton. 1992. Mineralogy and mineral weathering. Pages 378–425 in *Atmospheric Deposition and Forest Nutrient Cycling: A Synthesis of the Integrated Forest Study*. D. W. Johnson and S. E. Lindberg, editors. Springer-Verlag, New York, New York.

Arthur, M. A., G. B. Coltharp, and D. L. Brown. 1998. Effects of best management practices on forest streamwater quality in eastern Kentucky. *Journal of the American Water Resources Association* 34: 481–495.

Boring, L. R., W. T. Swank, and C. D. Monk. 1988. Dynamics of early successional forest structure in the Coweeta basin. Pages 161–179 in *Forest Hydrology and Ecology at Coweeta*. W. T. Swank and D. A. Crossley, Jr., editors. Springer-Verlag, New York, New York.

Breslow, N. E. 1996. Statistics in epidemiology: the case-control study. *Journal of the American Statistical Association* 91: 14–28.

Campbell, J. L., J. W. Hornbeck, W. H. McDowell, D. C. Buso, J. B. Shanley, and G. E. Likens. 1999. Dissolved organic nitrogen budgets for upland, forested ecosystems in New England. *Biogeochemistry* 49: 123–142.

Dai, K. H., C. E. Johnson, and C. T. Driscoll. 2001. Organic matter chemistry and dynamics in clear-cut and unmanaged hardwood forest ecosystems. *Biogeochemistry* 54: 51–83.

Douglass. J. E., and W.T. Swank. 1975. Effects of management practices on water quality and quantity: Coweeta Hydrologic Laboratory. North Carolina. Pages 1–13 in Proc. Municipal Watershed Management Symp., University Park, PA. 19-20 Sept. 1975. USDA, Forest Service General Tech. Rep. NE-13. USDA Forest Serv., Northeastern Forest Exp. Stn., Broomall, Pennsylvania.

Emmett, B. A., J. M. Anderson, and M. Hornung. 1990. The controls on dissolved nitrogen losses following two intensities of harvesting in a Sitka spruce forest (N. Wales). *Forest Ecology and Management* 41: 65–80.

Kalbitz, K., B. Glaser, and R. Bol. 2004. Clear-cutting of a Norway spruce stand: implications for controls on the dynamics of dissolved organic matter in the forest floor. *European Journal of Soil Science* 55: 401–413.

Kielland, K. 1994. Amino acid uptake by arctic plants. *Ecology* 75: 2373–2383.

Lewis, W. M., Jr., J. M. Melack, W. H. McDowell, M. McClain, and J. E. Richey. 1999. Nitrogen yields from undisturbed watersheds in the Americas. *Biogeochemistry* 46: 149–162.

Lewis, W. M., Jr. 2002. Yield of nitrogen from minimally disturbed watersheds of the United States. *Biogeochemistry* 57: 375–385.

Likens, G. E., and F. H. Bormann. 1995. *Biogeochemistry of a Forested Ecosystem*. Springer-Verlag, New York, New York.

Marks, P. L. 1974. The role of pin cherry in the maintenance of stability in Northern hardwood forest ecosystems. *Ecological Monographs* 44: 73–88.

Mattson, K. G., W. T. Swank, and J. B. Waide. 1987. Decomposition of woody debris in a regenerating, clear-cut forest in the Southern Appalachians. *Canadian Journal of Forest Research* 17: 712–720.

McDowell, W. H., and G. E. Likens. 1988. Origin, composition, and flux of dissolved organic matter in the Hubbard Brook valley. *Ecological Monographs* 58: 177–195.

McDowell, W. H., and T. Wood. 1984. Podzolization: soil processes control dissolved organic carbon concentrations in stream water. *Soil Science* 137: 23–32.

McGinty, D. T. 1976. Comparative root and soil dynamics on a white pine watershed in the hardwood forest in the Coweeta basin. PhD dissertation. University of Georgia, Athens.

Meyer, J. L., and C. M. Tate. 1983. The effects of watershed disturbance on dissolved organic carbon dynamics of a stream. *Ecology* 64: 33–44.

Michalzik, B., K. Kalbitz, J. H. Park, S. Solinger, and E. Matzner. 2001. Fluxes and concentrations of dissolved organic carbon and nitrogen: a synthesis for temperate forests. *Biogeochemistry* 52: 173–205.

Monk, C.D. 1975. Nutrient losses in particulate form as weir pond sediments from four unit watersheds in the Southern Appalachians. Pages 862–867 in *Mineral Cycling in Southeastern Ecosystems*. F. G. Howell, J. B. Gentry, and M. H. Smith, editors. US Energy Research Development Administration, Washington DC.

Monk, C. D., and F. P. Day. 1988. Biomass, primary production, and selected nutrient budgets for an undisturbed watershed. Pages 151–159 in *Forest Hydrology and Ecology at Coweeta*. W. T. Swank and D. A. Crossley, Jr., editors. Springer-Verlag, New York, New York.

Nelson, P. N., E. Cotsaris, and J. M. Oades. 1996. Nitrogen, phosphorus, and organic carbon in streams draining two grazed catchments. *Journal of Environmental Quality* 25: 1221–1229.

Olson, J. S. 1958. Rates of succession and soil changes on southern Lake Michigan sand dunes. *Botanical Gazette* 119: 123–170.

Perakis, S. S., and L. O. Hedin. 2007. State factor relationships of dissolved organic carbon and nitrogen losses from unpolluted temperate forest watersheds. *Journal of Geophysical Research: Biogeosciences* 112. doi:10.1029/2006JG000276.

Piirainen, S., L. Finer, H. Mannerkoski, and M. Starr. 2002. Effects of forest clear-cutting on the carbon and nitrogen fluxes through podzolic soil horizons. *Plant and Soil* 239: 301–311.

Qualls R. G. 2000. A comparison of the behavior of soluble organic and inorganic nutrients in forest ecosystems. *Forest Ecology and Management* 138: 29–50.

Qualls, R. G., and B. L. Haines. 1991. Geochemistry of dissolved organic nutrients in water percolating through a forest ecosystem. *Soil Science Society of America Journal* 52: 1112–1123.

Qualls, R. G., and B. L. Haines. 1992a. Measuring adsorption isotherms using continuous, unsaturated flow through intact soil cores. *Soil Science Society of America Journal* 56: 456–460.

Qualls, R. G., and B. L. Haines. 1992b. Biodegradability of dissolved organic matter in forest throughfall, soil-solution, and streamwater. *Soil Science Society of America Journal* 56: 578–586.

Qualls, R. G., B. L. Haines, and W. T. Swank. 1991. Fluxes of dissolved organic nutrients and humic substances in a deciduous forest ecosystem. *Ecology* 72: 254–266.

Qualls R. G., B. L. Haines, W. T. Swank, and S. W. Tyler. 2000. Fluxes of soluble organic and inorganic nutrients in clearcut and mature forests. *Soil Science Society of America Journal* 64: 1068–1077.

Qualls, R. G., B. L. Haines, W. T. Swank, and S. W. Tyler. 2002. Retention of dissolved organic nutrients by a forested ecosystem. *Biogeochemistry* 61: 135–171.

Qualls, R. G., and C. J. Richardson. 2003. Factors controlling concentration and export of dissolved organic nutrients in the Florida Everglades. *Biogeochemistry* 62: 197–229.

Ragsdale, H. L., and C.W. Berish. 1988. Trace metals in the atmosphere, forest floor, soil, and vegetation. Pages 367–382 in *Forest Hydrology and Ecology at Coweeta*. W. T. Swank and D. A. Crossley, Jr., editors. Springer-Verlag, New York, New York.

Ring, E. 1995. Nitrogen leaching before and after clear-felling of fertilized experimental plots in a *Pinus sylvestris* stand in central Sweden. *Forest Ecology and Management* 72: 151–166.

Seely B., K. Lajtha, and G. D. Salvucci. 1998. Transformation and retention of nitrogen in a coastal forest ecosystem. *Biogeochemistry* 42: 325–343.

Sollins, P., and F. M. McCorison. 1981. Nitrogen and carbon solution chemistry of old-growth coniferous forest watershed before and after cutting. *Water Resources Research* 17: 1409–1418.

Stevens, P., and M. Hornung. 1990. Effect of harvest intensity and groundflora establishment on inorganic N leaching from a Sitka Spruce plantation in N. Wales, U.K. *Biogeochemistry* 10: 53–65.

Smolander A., V. Kitunen, and E. Mälkönen. 2001. Dissolved soil organic nitrogen and carbon in a Norway spruce stand and an adjacent forest. *Biology and Fertility of Soils* 33: 190–196.

Swank, W. T., J. M. Vose, and K. J. Elliott. 2001. Long-term hydrologic and water quality responses following commercial clearcutting of mixed hardwoods on a southern Appalachian catchment. *Forest Ecology and Management* 143: 163–178.

Swank, W. T., and J. B. Waide. 1988. Baseline precipitation and stream chemistry. Pages 57–79 in *Forest Hydrology and Ecology at Coweeta*. W. T. Swank and D. A. Crossley, Jr., editors. Springer-Verlag, New York, New York.

Swank, W. T. 1988. Stream chemistry responses to disturbance. Pages 339–357 in *Forest Hydrology and Ecology at Coweeta*. W. T. Swank and D. A. Crossley, Jr., editors. Springer-Verlag, New York, New York.

Swank, W. T., L. W. Swift, and J. E. Douglass. 1988. Streamflow changes associated with forest cutting, species conversions, and natural disturbances. Pages 297–312 in *Forest Hydrology and Ecology at Coweeta*. W. T. Swank and D. A. Crossley, Jr., editors. Springer-Verlag, New York, New York.

Tate, C. M., and J. L. Meyer. 1983. Influence of hydrologic conditions and successional state on dissolved organic carbon export from forested watersheds. *Ecology* 64: 25–32.

Uselman, S. M., R. G. Qualls, and J. Lilienfein. 2007. Contribution of root- vs. leaf-litter to dissolved organic carbon leaching through soil. *Soil Science Society of America Journal* 71: 1555–1563.

Vitousek, P. M., J. R. Gosz, C. C. Grier, J. M. Melillo, W. A. Reiners, and R. L. Todd. 1979. Nitrate losses from disturbed ecosystems. *Science* 204: 469–474.

Walbridge, M. R., C. J. Richardson, and W. T. Swank. 1991. Vertical distribution of biological and geochemical phosphorus subcycles in two southern Appalachian forest soils. *Biogeochemistry* 13: 61–85.

Wood, T., F. H. Bormann, and G. K. Vogt. 1984. Phosphorus cycling in a northern hardwood forest: biological and chemical control. *Science* 223: 391–393.

6

Dynamics of Dissolved Organic Carbon in a Stream during a Quarter Century of Forest Succession

Judy L. Meyer*
Jackson R. Webster
Jennifer D. Knoepp
E. F. Benfield

Introduction

Dissolved organic carbon (DOC) is a heterogeneous mixture of compounds that makes up a large fraction of the organic matter transported in streams (Webster and Meyer 1997). It plays a significant role in many ecosystems. Riverine DOC links organic carbon cycles of continental and oceanic ecosystems. It is a significant trophic resource in stream food webs (e.g., Hall and Meyer 1998). DOC imparts color to lakes, regulating the amount of ultraviolet radiation reaching lake biota and influencing lake thermal regimes (Schindler and Curtis 1997). Cycling and biotic impact of metals are influenced by DOC concentration; for example, the concentration of methyl mercury in lakes and in lake biota increases with their DOC content (Driscoll et al. 1995). A synthesis of lake research identified colored DOC as a key characteristic of lakes, determining a lake's response to multiple anthropogenic stressors (Williamson et al. 1999). Because DOC regulates so many aspects of aquatic ecosystems, it is important to understand how natural and anthropogenic changes can alter its concentration in lotic ecosystems.

DOC transport in streams varies with watershed topography, extent of wetlands, soil type, nature of vegetation, fire history, hydrology, and land use. DOC concentration and transport increase as proportion of the watershed covered by wetlands or peat increases, and as soil carbon (C) content increases (e.g., Aitkenhead et al. 1999). Differences in soils' capacity for C sorption results in differences in stream DOC content (Nelson et al. 1993; see also Qualls et al., chapter 5, this volume). DOC export is less in watersheds with lower soil pH (Brooks et al. 1999). Forest type is an important predictor of DOC delivery to Adirondack lakes (Canham et al.

* Corresponding author: 498 Shoreland Dr., Lopez Island WA 98261 USA

2004). Increases in DOC delivery to lakes after extensive forest fires have been attributed to altered hydrology (Schindler et al. 1997). Human alterations of watersheds for agriculture, forest harvest, or development also influences DOC transport in streams (e.g., Eckhardt and Moore 1990).

The impact of forest-management practices, such as clearcutting, on stream DOC concentration and export is not the same in all landscapes. Clearcutting watersheds with extensive peatlands has resulted in elevated DOC concentration and export (Cummins and Farrell 2003; Nieminen 2004). Higher DOC concentration and export have also been observed after cutting in areas where extensive logging slash was left in the channel (Moore 1989). Clearcutting conifers in Oregon (Dahm 1980), hardwoods in New Hampshire (Johnson et al. 1995), and boreal forests in Canada (France et al. 2000) resulted in elevated DOC export as well as elevated DOC concentrations in downstream lakes (Carignan et al. 2000; Lamontagne et al. 2000). Elevated export has been observed for periods as long as 15 years after forest cutting (Johnson et al. 1995), although the chemical composition of DOC in streamwater was not altered by cutting (Dai et al. 2001). Selective cutting has also resulted in slight increases in DOC export, largely because of increased water yield (Kreutzweiser et al. 2004). In contrast, other studies of clearcuts have found either no difference in DOC export (Grieve 1994) or reduced export several years after cutting (Moore 1989). Previous studies at Coweeta Hydrologic Laboratory have observed decreases in streamwater DOC concentration and export after forest removal (Meyer and Tate 1983; Meyer et al. 1988). Experimental and hurricane-induced removal of riparian vegetation at Coweeta also resulted in lower concentrations of DOC in soil water (Yeakley et al. 2003). Coweeta streams with a history of forest removal had lower DOC concentration and export than reference streams when sampled 10–20 years after treatment, although differences resulting from forest practices were less than those resulting from changes in hydrology (Tate and Meyer 1983). In contrast, Qualls et al. (see chapter 5, this volume) observed elevated DOC concentrations and flux from Coweeta soil horizons during the first two years after the experimental clearcutting of an area on Watershed (WS) 2.

DOC concentration and transport in Big Hurricane Branch, the stream draining the clearcut watershed that is the subject of this book, were less than in the reference stream (Hugh White Creek) two years after the cut (Meyer and Tate 1983). These differences were attributed to lower DOC concentration in the subsurface water entering the channel and to reduced in-stream generation of DOC (Meyer and Tate 1983). Lower DOC concentrations and transport continued for three more years, although DOC concentration in Big Hurricane Branch was increasing relative to the reference (Meyer et al. 1988). Based on five years of data from the first seven years after cutting, Meyer et al. (1988) predicted that DOC concentrations in Big Hurricane Branch would continue to increase until they were indistinguishable from those in the reference stream.

We now have an additional 20 years of data on DOC concentration in these two streams. One objective of this study was to determine whether the prediction based on those initial five years of data was correct, that is, whether DOC concentrations in Big Hurricane Branch became more similar to those in the reference stream as the forest on the clearcut watershed recovered. A second objective was to describe seasonal and interannual patterns of DOC concentration and transport in the two streams. The final objective was to explore potential causes of the seasonal and interannual patterns observed.

Methods

Big Hurricane Branch drains WS 7, which was clearcut using cable logging in 1977 (see chapter 1 for a more detailed description of this operation). Hugh White Creek drains WS 14 (figure 6.1), one of the long-term reference watersheds at Coweeta (Webster et al. 1997). The two streams are similar in watershed size, discharge, gradient, length, and elevation, but they differ in aspect; WS 7 faces south; whereas WS 14 faces north (see Webster et al., chapter 10, this volume). We began collecting water samples from the two streams for DOC analysis in 1979, two years after the cut; we did not have DOC analytical capacity prior to that time. Samples were collected at two-week intervals for the first year of the study and weekly after that. In this chapter, we consider the data from July 1979 through July 2004.

Water samples were collected immediately above the weir pond in each stream, filtered through ashed glass fiber filters (Gelman A/E, nominal pore size 0.3 μm), and refrigerated for one to four weeks until analysis. From 1979 through 2000, samples were analyzed using a Dohrmann Envirotech DC-54 carbon analyzer or an Oceanography International Organic Carbon analyzer, both of which use persulfate oxidation. In 2001, we began analyzing samples using a Shimadzu TOC-5000A Total Organic Carbon analyzer, which uses high-temperature combustion in the

Figure 6.1   Upstream view of a 120° V-notch weir installation on Coweeta WS 14, October 2006. (Photo by J. Webster)

presence of a catalyst. Based on one year of samples analyzed using both methods, we determined that PS = 0.196 + 0.731 HT (r^2 = 0.78), where PS is DOC concentration measured using persulfate oxidation, and HT is DOC concentration measured using high-temperature combustion. Because the longest record of DOC concentration was based on persulfate oxidation, all concentrations in this chapter have been converted to that method using this equation.

Mean DOC concentrations were calculated by month, season, and water year. The seasons were defined as follows: January–March is winter; April–June is spring; July–September is summer; and October–December is autumn. Water years begin November 1 and end October 31; for example, the 1980 water year is November 1, 1980, through October 31, 1981.

Gage height recorded at the time of sampling was used to calculate instantaneous discharge (L/s), which we combined with measured concentration to determine daily DOC transport. Monthly, seasonal, and annual means for discharge used in regressions were obtained from the US Forest Service's long-term record of discharge from these two watersheds.

In this chapter, we relate DOC concentrations to leaf standing crop in the stream benthos (benthic leaf litter) and soil organic C content. Benthic leaf litter was measured in both streams seasonally in 1985–1986 (Golladay et al. 1989); monthly, from August 1986–August 1987 (Stout et al. 1993); and every other month, from November 1993 to September 1994 (see Webster et al., chapter 10, this volume). Soil organic C content at depths of 0–10 cm and 10–30 cm was measured annually on WS 7 during 1979–1985, 1992–1994, and 1998 using methods described by Knoepp et al., in chapter 4 of this volume. Measures of soil organic C content are not available for WS 14, the reference stream's watershed.

Results

Both streams exhibited a consistent seasonal pattern in DOC concentration (figure 6.2). Concentrations were highest in autumn, declined through winter, reached a nadir in March, and increased through summer. Average DOC concentrations were higher in the reference stream in all months, and seasonal concentration excursions were also greater in this stream. The mean ratio of DOC concentration in Big Hurricane Branch to DOC concentration in the reference stream was less than one in all months (figure 6.2). The ratio had a seasonal pattern, with ratios ranging from 0.75 to 0.85 in December through April, hovering around 0.65 from May through September, and increasing in autumn. Hence, the difference in concentration between the two streams was greatest in the growing season and least in the dormant season.

The range in mean annual DOC concentration over the 25-year record was the same as the range in average monthly concentrations (figure 6.3 cf. figure 6.2). The temporal pattern of mean annual DOC concentration over this quarter century was similar in the two streams (figure 6.3), and mean annual DOC concentrations in the two streams were highly correlated ($r = 0.86$, $P < 0.0001$). Average annual concentrations in both streams increased initially, then decreased, with highest

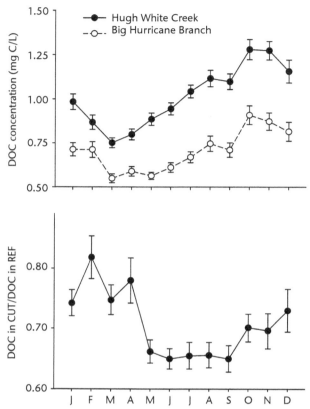

Figure 6.2 *Upper panel*: Mean (±SE) monthly DOC concentration in Big Hurricane Branch (clearcut watershed) and Hugh White Creek (reference watershed) over the 25-yr record. *Lower panel*: Monthly mean (±SE) of the ratio of DOC concentration in Big Hurricane Branch (CUT) to DOC concentration in Hugh White Creek (REF) over the 25-yr record.

concentrations 10–15 years after the cut (1988–1993) (figure 6.3). The annual mean of the ratio of DOC concentration in Big Hurricane Branch to DOC concentration in the reference stream showed a similar pattern (figure 6.3), indicating that concentrations initially converged and then diverged.

The pattern of DOC concentration was not simply a result of changing hydrologic conditions. Mean annual DOC concentration was not correlated with mean annual discharge in either stream ($r = 0.14, P = 0.50$ in Big Hurricane Branch; $r = 0.06, P = 0.77$ in reference). However, changing hydrologic conditions did impact the ratio of DOC concentration in the two streams. Multiple regression analysis ($R^2 = 0.35, P = 0.01$) revealed that the ratio of DOC concentration in Big Hurricane Branch to DOC concentration in the reference stream increased with increasing discharge ($P = 0.02$) and decreased with time since the cut ($P = 0.04$), although it was always less than one. Mean annual discharge in the reference stream was used

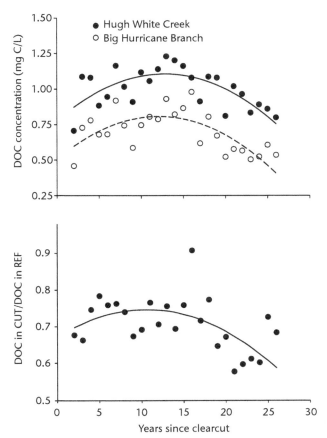

Figure 6.3 *Upper panel*: Mean annual DOC concentration in Big Hurricane Branch (clearcut watershed) and Hugh White Creek (reference watershed) plotted versus years since WS 7 was clearcut (1977). Solid line is the regression $y = 0.778 + 0.051 \, x - 0.002 \, x^2$, $r^2 = 0.43, P = 0.003$. Dotted line is the regression $y = 0.506 + 0.050 \, x - 0.002 \, x^2$, $r^2 = 0.47$, $P = 0.001$. *Lower panel*: Annual mean of the ratio of DOC concentration in Big Hurricane Branch (CUT) to DOC concentration in Hugh White Creek (REF) versus years since WS 7 was clearcut. Line is the regression $y = 0.672 + 0.14 \, x - 0.0007 \, x^2$, $r^2 = 0.29, P = 0.03$.

in this analysis as an indicator of hydrologic conditions; it was not related to time since the cut ($P = 0.9$). This analysis indicates that differences in DOC concentrations in the two streams were greatest (i.e., the ratio was farthest from one) in dry years and as forest succession proceeded.

The long-term temporal pattern of the ratio in concentrations differed by season (figure 6.4). The ratio was most variable in autumn but exhibited no consistent trend over the quarter century. In contrast, during both winter and summer, ratios increased in the first three to five years, followed by a decline. Ratios in spring increased for the first 17 years, but have declined since then. Some of this variation in seasonal ratios can be explained by mean seasonal discharge in the reference

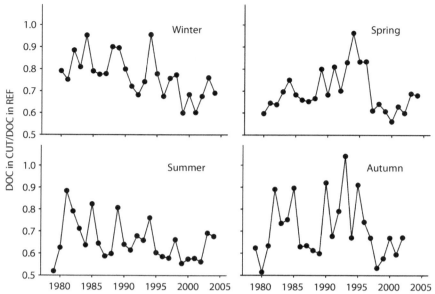

Figure 6.4   Seasonal mean of the ratio of DOC concentration in Big Hurricane Branch (CUT) to DOC concentration in Hugh White Creek (REF) for each season and year of record.

stream across years ($r^2 = 0.12$, $P = 0.0005$). Seasonal average DOC concentration declined with increasing discharge in both streams, but the decline was greater in the reference watershed; therefore the ratio of DOC concentration in Big Hurricane Branch to DOC concentration in the reference stream increased with increasing discharge. A multiple regression of seasonal ratios versus seasonal discharge in the reference stream and time since cut ($R^2 = 0.17$, $P = 0.0002$) is consistent with what was described for the annual data in the previous paragraph: the ratio remained less than one but increased with discharge ($P = 0.0004$) and decreased as forest succession proceeded ($P = 0.03$). Seasonal differences in DOC concentration in the two streams were greatest during dry seasons and as forest succession proceeded.

Some of the differences in DOC concentration over seasons and years can be explained by changes in amount of benthic leaf litter in the streams. The amount of benthic leaf litter explained more of the variation in seasonal DOC concentration in Big Hurricane Branch ($r^2 = 0.43$) than in the reference stream ($r^2 = 0.18$). When data from both streams were combined, the amount of benthic leaf litter explained 35% of the variation in seasonal average DOC concentration (figure 6.5). Including seasonal discharge did not improve this relationship.

These findings can be compared with data from two other Coweeta streams in which benthic leaf litter was altered experimentally by excluding leaf litter in one stream while keeping the other as a reference (Meyer et al. 1998). A regression of annual mean DOC concentration in the two reference streams, the clearcut stream, and the litter-excluded stream versus mean annual benthic leaf litter explained 34% of the variation in DOC concentration among streams and years (figure 6.6). These

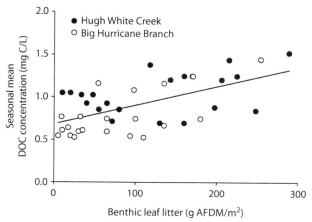

Figure 6.5 Seasonal mean DOC concentration in Hugh White Creek (reference watershed) and Big Hurricane Branch (clearcut watershed) versus benthic leaf litter in the stream. Line is the regression y = 0.683 + 0.0022 x, r^2 = 0.35, $P < 0.0001$.

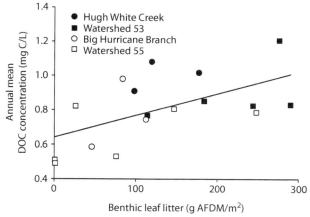

Figure 6.6 Annual mean DOC concentration in Hugh White Creek (reference watershed), Watershed 53 (litter exclusion reference stream), Big Hurricane Branch (clearcut watershed), and Watershed 55 (litter-excluded stream) plotted versus mean annual benthic leaf litter. Line is the regression y = 0.642 + 0.0013 x, r^2 = 0.34, $P = 0.01$.

data strongly suggest that leaching of benthic leaf litter is a significant source of DOC in these headwater streams.

Leaching of soil organic matter is another source of DOC to streams. It is likely that this DOC source varies as organic matter in the soil changes with forest succession and as the amount of water passing through the soil changes. We have expressed this potential source as annual average soil organic C in the top 10 cm divided by seasonal discharge (figure 6.7). Seasonal mean concentrations of DOC in Big

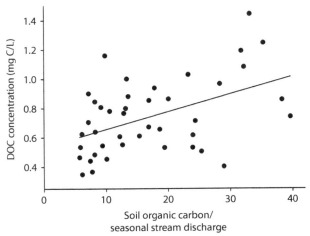

Figure 6.7   Seasonal mean DOC concentration in Big Hurricane Branch plotted versus the ratio of soil organic C content in the top 10 cm / seasonal discharge. Regression is y = 0.535 + 0.012 x, $r^2 = 0.22$, $P = 0.002$.

Hurricane Branch increased as this source increased, and the regression explained 22% of the variation in seasonal DOC concentration (figure 6.7). Seasonal mean DOC concentrations were also significantly correlated with soil organic C from 10–30 cm divided by seasonal discharge ($P = 0.01$), but that regression explained somewhat less of the variation in seasonal DOC concentration ($r^2 = 0.15$). Including only DOC data from seasons with measures of both soil organic C and benthic leaf litter reduces the number of data points that could be used in a multiple regression from 42 to 8, and the two measures are correlated ($P = 0.03$). Therefore, we cannot use multiple regression analysis to determine if combining in-stream and watershed sources would better predict stream DOC concentrations than either variable alone.

Water samples were collected at weekly intervals without intensive sampling during storms. DOC concentration increases during storms in these streams (Meyer and Tate 1983), so these data have limited usefulness in determining annual transport of DOC from the watersheds. Multiplying concentration by measured discharge at the time of sampling does provide an instantaneous measure of DOC transport. We have calculated an annual average of these values, which we call the *mean daily DOC load*, but because it does not include systematic samples of storms, it is probably an underestimate of the true load. Mean daily DOC load was consistently higher in the reference stream than in Big Hurricane Branch, with a temporal pattern similar to that observed with DOC concentration (figure 6.8 cf. figure 6.3), though with greater variability, so regressions were not significant. Annual mean ratio of load in Big Hurricane Branch to load in the reference stream declined with time since cutting (figure 6.8). Including mean annual discharge in a multiple regression did not improve the fit. The mean daily load of DOC transported from the clearcut watershed relative to mean daily load transported from the reference watershed declined with forest regrowth.

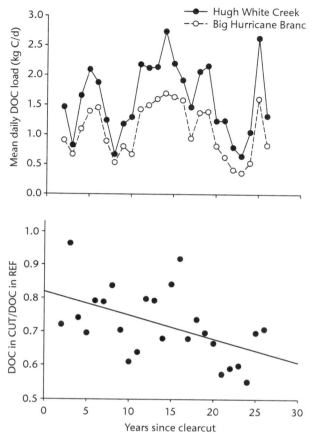

Figure 6.8   *Upper panel*: Mean daily DOC load (kg C/d) in Hugh White Creek (reference watershed) and Big Hurricane Branch (clearcut watershed) plotted versus years since WS 7 was clearcut. *Lower panel*: Ratio of mean daily DOC load in Big Hurricane Branch to mean daily DOC load in Hugh White Creek plotted versus years since WS 7 was clearcut. Line is the regression $y = 0.820 - 0.007\,x$, $r^2 = 0.25$, $P = 0.01$.

Discussion

An analysis of the first five years of data in this study led to a prediction that DOC concentrations would increase in Big Hurricane Branch, so that over time, DOC concentrations would be similar in the two streams (Meyer et al. 1988). DOC concentration increased initially in Big Hurricane Branch relative to the reference stream, but this trend was not sustained (figure 6.3). Therefore, concentrations in the two streams did not become more similar as the forest grew after cutting. In fact, during winter and summer, DOC concentrations in the two streams became more different over time (figure 6.4). Analyses of both annual and seasonal mean ratios of DOC concentration in Big Hurricane Branch to DOC concentration in the reference stream indicated that DOC concentrations in the two streams were least similar during dry periods and later in forest succession. Hence, analyses of an additional 20 years of data showed that the prediction based on only the first five years of data was not correct. Long-term trends cannot reliably be predicted from short-term data sets.

Worrall et al. (2005) measured inorganic C flux from weekly samples taken in Big Hurricane Branch and used estimates of DOC and particulate organic carbon (POC) flux to calculate total C flux from reference and cut watersheds. We have revised their estimate of total C flux using more recent data for organic C (table 6.1). Both DOC and POC fluxes are higher than reported by Worrall et al. (2005), so total C flux estimates are also somewhat higher. Total C flux from the clearcut watershed is somewhat lower than from the reference watershed, but the range of flux estimates overlap (table 6.1). DOC remains a relatively small component (7%–8%) of total C flux, which is dominated by inorganic C flux (Worrall et al. 2005). For these estimates, POC flux exceeded DOC flux, which is somewhat unusual. However, this may be a result of sampling technique. POC sampling included intensive storm-water collection; whereas DOC calculations are based on weekly grab samples.

Observed DOC concentrations are the net result of DOC supply and in-stream DOC consumption. Limited measurements of in-stream DOC uptake showed little difference between Big Hurricane Branch and the reference stream when labile DOC was added to both streams two years after clearcutting (Meyer et al. 1988). This suggests that differences in rate of supply may be more responsible for the differences reported here, although we do not know how uptake has changed over time. As Webster et al. report in chapter 10 of this volume, leaf litter inputs to Big Hurricane Branch were lower immediately after the cut; but input amount was nearly the same as in the reference stream within five years. However, the quality of the inputs differed in that successional species having less refractory litter dominated the clearcut watershed. Because of the differences in litter quality, stream

Table 6.1   Total fluvial carbon flux (t km^{-2} yr^{-1}) from the reference and clearcut watersheds.

	Reference watershed		Clearcut watershed		Source
	Mean	Range	Mean	Range	
Dissolved CO_2	3.72	2.33–13.94	3.48	1.68–14.7	1
DIC	4.25	1.0–11.0	2.89	1.0–10.5	1
DOC	0.99	0.40–1.64	0.69	0.26–1.05	2
POC	2.9	—	2.0	—	3
Dissolved CH_4	0.26	0.12–0.40	0.26	0.12–0.40	1
Total	12.12	6.75–29.9	9.32	5.06–28.7	

Note: This table is an update of table 1 of Worrall et al. (2005) using more recent DOC and POC data.
[1] Worrall et al. (2005), 1979–1999 on WS 2 (reference) and WS 7 (clearcut).
[2] This chapter, using data from 1979–1999 from WS 14 (reference) and WS 7 (clearcut).
[3] Webster et al. (1990), assuming POM is 50% C and using data from 1984–1985 from WS 7 (clearcut) and an average of WS 14 and 18 (both reference). These POC data include storm sampling; whereas the others are from weekly samples.

nutrient concentration, and stream temperature, leaf litter decayed more rapidly in Big Hurricane Branch so that little remained by late spring and summer. In addition, leaf litter could more easily be washed downstream in Big Hurricane Branch because there was less wood in the channel. The seasonal pattern of DOC concentration in the two streams (figure 6.2) was consistent with these differences in litter quality, decay rate, and retention in debris dams. DOC concentrations were the most similar in autumn and winter, but they diverged in spring and summer, when most of the litter inputs had decayed or been washed downstream in Big Hurricane Branch.

The bioavailability of DOC leached from riparian leaf litter differs by species, and the leachate from early successional species is more bioavailabile (McArthur and Richardson 2002). Although uptake rates measured using the same labile DOC source were similar in the two streams shortly after clearcutting (Meyer et al. 1988), we have no data on uptake rates for DOC naturally entering each stream. Since leachate from early successional leaves falling into Big Hurricane Branch was likely more bioavailable, the observed lower ambient DOC concentrations are what would be expected in that stream. The reference stream showed consistent downstream increases in DOC concentration; whereas downstream concentrations did not change in Big Hurricane Branch two years after the cut (Meyer and Tate 1983). This observation is consistent with more rapid removal of DOC in Big Hurricane Branch, although we do not know if these differences in longitudinal patterns have been sustained over the years in these streams.

The absence of a longitudinal increase in DOC concentration in Big Hurricane Branch is also consistent with the lower rates of DOC supply from the clearcut watershed. The correlation of DOC concentration with both benthic leaf litter (figure 6.5) and soil organic C content/seasonal discharge (figure 6.7) provides even stronger evidence for differences in both in-stream and watershed sources of DOC to the two streams. The importance of in-stream DOC sources is not unique to these two streams. Mean annual DOC concentration in four very different Coweeta streams can be predicted by benthic leaf litter (figure 6.6) despite the fact that watershed size (~60 ha vs. ~6 ha), discharge (~20 L/s vs. < 2 L/s), and streambed area differ by an order of magnitude. The amount of water flowing over and leaching DOC from benthic leaf litter is similar per m^2 in these streams, so DOC concentration is related to benthic leaf litter expressed per m^2. Data from other southern Appalachian streams show a similar pattern of autumn peaks in DOC concentration and evidence of significant in-stream generation of DOC (Mulholland and Hill 1997).

Organic C reservoirs in the soil have been identified as a significant source of DOC in streams and lakes in many settings. Differences in soil organic C storage explained 91% of the variance in annual streamwater DOC fluxes among 17 British rivers (Hope et al. 1997). Soil organic C pools correlated with mean DOC concentrations in Scottish watersheds of different sizes, with the strongest relations in the small (< 5 km^2) watersheds (Aitkenhead et al. 1999). Predictive models of stream DOC export have been developed in which a terrestrial reservoir of soil DOC builds up during low-flow periods and then flushes when the water table rises into this reservoir (Boyer et al. 1996). The relationship between the DOC concentration in Big Hurricane Branch and the organic C content of the top 10 cm of soil in its watershed (figure 6.7) is consistent with these findings from other ecosystems.

A variable fraction of DOC leached from organic matter in the upper soil horizons is sorbed in the mineral soil, leading many researchers to conclude that sorption and hydrology are the main regulators of DOC losses from terrestrial ecosystems (e.g., Neff and Asner 2001; see also Qualls et al., chapter 5, this volume). High levels of DOC adsorption in the mineral soil at Coweeta have been attributed to the soil's high content of Fe and Al oxyhydroxides (see Qualls et al. chapter 5, this volume). Although we do not have data on the Fe and Al oxyhydroxide content of soils from the reference or from the clearcut watershed, we know that the Fannin soil series is a dominant on the clearcut watershed and that Trimont is a dominant on the reference watershed (Thomas 1996; see Knoepp et al., chapter 4, this volume), Analyses from soil pits characterizing these soil series reveal Fe and Al oxyhydroxide content of Fannin series soils to be about a third of the Fe and Al oxyhydroxide content of Trimont series soils (USDA 2005). Hence the lower DOC concentrations in Big Hurricane Branch cannot be explained by increased DOC sorption in its mineral soils because the Fe and Al oxyhydroxide content of its soils is less than in the soils from Hugh White Creek's watershed.

A 25–year record of DOC concentration in stream water does not exist for many other streams. Two larger British rivers have long records (30–40 years) during which DOC concentration increased 1 mg/L and 3 mg/L (Worrall and Burt 2004). These increases were attributed to increasing temperatures and drought, resulting in accelerated decomposition of peat via release from inhibition by phenolics (Worrall and Burt 2004). A 20–year record showed a 50% decline in DOC export to Canadian lakes as a result of drought-induced forest fires and, especially, lower water yield (Schindler et al. 1997). The long-term pattern of DOC concentration and export in Coweeta streams showed neither sustained increase nor decrease. Mean annual DOC concentration was not simply a function of stream discharge but also changed as a function of the amount of leachable organic matter in both the stream and the soils of its watershed. As these stores of leachable organic matter changed during forest succession, so did DOC in streamwater. The long-term trend of declining mean daily DOC load in Big Hurricane Branch relative to the reference stream (figure 6.8) suggests that there has been a depletion of DOC sources during the first quarter century of forest recovery from clearcutting.

Acknowledgments

One does not collect 25 years of data without the assistance of an army of helpers—to whom we are very grateful. The following are some of those whom we thank: Cathy Tate for field and lab assistance at the beginning of the study, James Buchanan (Jim Buck) and Robert McCollum for weekly sample collection, Barbara Reynolds and James Deal for supervising the filtration of DOC samples through the years, Sue Eggert for assistance with sample transport, and Rebecca Auxier and

Tom Mattox for DOC analyses. This research was funded by several grants from the National Science Foundation Long-term Ecological Research Program.

Literature Cited

Aitkenhead, J. A., D. Hope, and M. F. Billett. 1999. The relationship between dissolved organic carbon in stream water and soil organic carbon pools at different spatial scales. *Hydrological Processes* 13: 1289–1302.

Boyer, E. W., G. M. Hornberger, K. E. Bencala, and D. M. McKnight. 1996. Overview of a simple model describing variation of dissolved organic carbon in an upland catchment. *Ecological Modelling* 86: 183–188.

Brooks, F. D., D. M. McKnight, and K. E. Bencala. 1999. The relationship between soil heterotrophic activity, soil dissolved organic carbon (DOC) leachate, and catchment-scale DOC export in headwater catchments. *Water Resources Research* 35: 1895–1902.

Canham, C. D., M. L. Pace, M. J. Papaik, A. G. Primack, R. J. Maranger, R. P. Curran, and D. M. Spada. 2004. A spatially explicit watershed-scale analysis of dissolved organic carbon in Adirondack lakes. *Ecological Applications* 14: 839–854.

Carignan, R., P. D'Arcy, and S. Lamontagne. 2000. Comparative impacts of fire and forest harvesting on water quality in Boreal Shield lakes. *Canadian Journal of Fisheries and Aquatic Sciences* 57 (Suppl. 2): 105–117.

Cummins, T., and E. P. Farrell. 2003. Biogeochemical impacts of clearfelling and reforestation on blanket-peatland streams II. Major ions and dissolved organic carbon. *Forest Ecology and Management* 180: 557–570.

Dahm, C. N. 1980. Studies on the distribution and fates of dissolved organic carbon. Dissertation. Oregon State University, Corvallis.

Dai, K. O. H., C. E. Johnson, and C. T. Driscoll. 2001. Organic matter chemistry and dynamics in clear-cut and unmanaged hardwood forest ecosystems. *Biogeochemistry* 54: 51–83.

Driscoll, C. T., V. Blette, C. Yan, C. L. Schofield, R. Munson and J. Holsapple. 1995. The role of dissolved organic carbon in the chemistry and bioavailability of mercury in remote Adirondack lakes. *Water, Air, and Soil Pollution* 80: 499–508.

Eckhardt, B. W., and T. R. Moore. 1990. Controls on dissolved organic carbon concentrations in streams, southern Quebec. *Canadian Journal of Fisheries and Aquatic Sciences* 47: 1537–1544.

France, R., R. Steedman, R. Lehman, and R. Peters. 2000. Landscape modification of DOC concentration in boreal lakes: implications for UV-B sensitivity. *Water, Air, and Soil Pollution* 122: 153–162.

Golladay, S. W., J. R. Webster, and E. F. Benfield. 1989. Changes in stream benthic organic matter following watershed disturbance. *Holarctic Ecology* 12: 96–105.

Grieve, I. 1994. Dissolved organic carbon dynamics in two streams draining forested catchments at Loch Ard, Scotland. *Hydrological Processes* 8: 457–464.

Hall, R. O., Jr., and J. L. Meyer. 1998. The trophic significance of bacteria in a detritus-based stream food web. *Ecology* 79: 1995–2012.

Hope, D., M. F. Billett, R. Milne, and T. A. W. Brown. 1997. Exports of organic carbon in British rivers. *Hydrological Processes* 11: 325–344.

Johnson, C. E., C. T. Driscoll, T. J. Fahey and T. G. Siccama. 1995. Carbon dynamics following clear-cutting of a northern hardwood forest. Pages 463–488 in *Carbon Forms and Functions in Forest Soils*. W. W. McFee, and J. M. Kelly, editors. Soil Science Society of America, Madison, Wisconsin.

Kreutzweiser, D. P., S. S. Capell, and F. D. Beall. 2004. Effects of selective forest harvesting on organic matter inputs and accumulation in headwater streams. *Northern Journal of Applied Forestry* 21: 19–30.

Lamontagne, S., R. Carignan, P. D'Arcy, Y. T. Prairie, and D. Pare. 2000. Element export in runoff from eastern Canadian Boreal Shield drainage basins following forest harvesting and wildfires. *Canadian Journal of Fisheries and Aquatic Sciences* 57 (Suppl. 2): 118–128.

McArthur, M. D., and J. S. Richardson. 2002. Microbial utilization of dissolved organic carbon leached from riparian litterfall. *Canadian Journal of Fisheries and Aquatic Sciences* 59: 1668–1676.

Meyer, J. L., and C. M. Tate. 1983. The effects of watershed disturbance on dissolved organic carbon dynamics of a stream. *Ecology* 64: 33–44.

Meyer, J. L., C. M. Tate, R. T. Edwards, and M. T. Crocker. 1988. The trophic significance of dissolved organic carbon in streams. Pages 269–278 in *Forest Hydrology and Ecology at Coweeta*. W. T. Swank, and D. A. Crossley, Jr., editors. Springer-Verlag, New York, New York.

Meyer, J. L., J. B. Wallace, and S. L. Eggert. 1998. Leaf litter as a source of dissolved organic carbon in streams. *Ecosystems* 1: 240–249.

Moore, T. R. 1989. Dynamics of dissolved organic carbon in forested and disturbed catchments, Westland, New Zealand. 1. Maimai. *Water Resources Research* 25: 1321–1330.

Mulholland, P. J., and W. R.Hill. 1997. Seasonal patterns in streamwater nutrient and dissolved organic carbon concentrations: separating catchment flow path and in-stream effects. *Water Resources Research* 33: 1297–1306.

Neff, J. C., and G. P. Asner. 2001. Dissolved organic carbon in terrestrial ecosystems: synthesis and a model. *Ecosystems* 4: 29–48.

Nelson, P. N., J. A. Baldock, and J. M. Oades. 1993. Concentration and composition of dissolved organic carbon in streams in relation to catchment soil properties. *Biogeochemistry* 19: 27–50.

Nieminen, M. T. 2004. Export of dissolved organic carbon, nitrogen and phosphorus following clear-cutting of three Norway spruce forests growing on drained peatlands in southern Finland. *Silva Fennica* 38: 123–132.

Schindler, D. W., P. J. Curtis, S. E. Bayley, B. R. Parker, K. G. Beaty, and M. P. Stainton. 1997. Climate-induced changes in the dissolved organic carbon budgets of boreal lakes. *Biogeochemistry* 36: 9–28.

Schindler, D. W., and P. J. Curtis. 1997. The role of DOC in protecting freshwaters subjected to climatic warming and acidification from UV exposure. *Biogeochemistry* 36: 1–8.

Stout, B. M. III, E. F. Benfield, and J. R. Webster. 1993. Effects of a forest disturbance on shredder production in southern Appalachian headwater streams. *Freshwater Biology* 29: 59–69.

Tate, C. M., and J. L. Meyer. 1983. The influence of hydrologic conditions and successional state on dissolved organic carbon export from forested watersheds. *Ecology* 64: 25–32.

Thomas, D.J., 1996. Soil Survey of Macon County, North Carolina. USDA Natural Resource Conservation Service, p. 322.

USDA, N.R.C.S., 2005. Web Soil Survey (version 2.1). USDA, Natural Resource Conservation Service.

Webster, J. R., S. W. Golladay, E. F. Benfield, D. J. D'Angelo, and G. T. Peters. 1990. Effects of forest disturbance on particulate organic matter budgets of small streams. *Journal of the North American Benthological Society* 9: 120–140.

Webster, J. R., S. W. Golladay, E. F. Benfield, J. L. Meyer, W. T. Swank, and J. B. Wallace. 1992. Catchment disturbance and stream response: an overview of stream research at

Coweeta Hydrologic Laboratory. Pages 231–253 in *River Conservation and Management*. P. J. Boon, P. Calow, and G. E. Petts, editors. John Wiley and Sons, London, England.

Webster, J. R., and J. L. Meyer (editors). 1997. Stream Organic Matter Budgets. Special issue of *Journal of the North American Benthological Society* 16: 3–161.

Webster, J. R., J. L. Meyer, J. B. Wallace, and E. F. Benfield. 1997. Organic matter dynamics in Hugh White Creek, Coweeta Hydrologic Laboratory, North Carolina, USA. Pages 74–78 in. J. R. Webster and J. L. Meyer, editors. *Stream Organic Matter Budgets*. Special issue of *Journal of the North American Benthological Society* 16: 3–161.

Williamson, C. E., D. P. Morris, M. L. Pace, and A. G. Olson. 1999. Dissolved organic carbon and nutrients as regulators of lake ecosystems: resurrection of a more integrated paradigm. *Limnology and Oceanography* 44: 795–803.

Worrall, F., and T. Burt. 2004. Time series analysis of long-term river dissolved organic carbon records. *Hydrological Processes* 18: 893–911.

Worrall, F., W. T. Swank, and T. Burt. 2005. Fluxes of inorganic carbon from two forested catchments in the Appalachian Mountains. *Hydrological Processes* 19: 3021–3035.

Yeakley, J. A., D. C. Coleman, B. L. Haines, B. D. Kloeppel, J. L.Meyer, W. T. Swank, B. W. Argo, J. M. Deal and S. F. Taylor. 2003. Hillslope nutrient dynamics following upland riparian vegetation disturbance. *Ecosystems* 6: 154–167.

7

Wood Decomposition Following Clearcutting at Coweeta Hydrologic Laboratory

Kim G. Mattson*
Wayne T. Swank

Introduction

Most of the forest on Watershed (WS) 7 was cut and left on site to decompose (figure 7.1). This chapter describes the rate and manner of wood decomposition and also quantifies the fluxes from decaying wood to the forest floor on WS 7. In doing so,we make the case that wood and its process of decomposition contributes to ecosystem stability. We also review some of the history of wood decomposition and place our results in the context of detrital organic matter pools on the watershed.

Much of our understanding of wood decay has come out of studies on how to prevent decay (reviewed by Hunt and Garratt 1938; Campbell 1952; Cartwright and Findlay 1958). Indeed, the prevention of wood decay is still a big business—for example, the Forest Products Laboratory of Madison, Wisconsin, estimates that "billions of dollars" are spent each year to replace wood products destroyed by wood decay fungi. We now also recognize the positive aspects of wood decay. With the promotion of an ecosystem perspective as a way to study nature (e.g., Odum 1969), decomposition began to be recognized as an important part of nutrient cycles and energy flow. The development of the ecosystem perspective helped to stimulate studies of the role of wood and its decomposition under natural conditions (McFee and Stone 1966; Swift 1977) and more studies of wood decomposition under natural conditions began to appear in the literature.

As reviews of wood decay began to appear (Spaulding and Hansbrough 1944; Aho 1974; Kaarik 1974; Swift 1977), it became apparent that reported wood decay rates were highly variable. The extensive review by Harmon et al. (1986) perhaps best demonstrates the wide variability in wood decay rates, where 26 reviewed studies of wood in situ (i.e., in the forests as opposed to laboratory studies) showed

* Corresponding author: Ecosystems Northwest, 189 Shasta Av., Mount Shasta, CA 96067 USA

Figure 7.1   Wood on WS 7 at year 1. Nearly 100 t/ha of wood was left on WS 7 following clearcutting. The volumes and densities were measured in 4 x 4 m plots during year 1 and provided the basis for remeasurements in years 6 and 11. (USDA Forest Service photo)

ranges of decay rates of more than 100-fold. It appeared difficult to draw meaningful quantitative relationships because even decay rates for the same species from the same region could vary by up to 25-fold. Most reviews have described specific details of various studies about the wood or site conditions and offered speculations on why decay rates may have been low or high in specific instances. But no relative comparison or ordering of the potential factors controlling wood decay had been attempted in most of these reviews, giving the impression that wood decay may be hopelessly variable. Meentemeyer (1984), however, developed a statistical model that explained 80% of the variation in decay of fine litter material over continental scales as a function of annual actual evapotranspiration (AET) combined with lignin content. This same analysis was performed for 67 reported wood decay rates (weight loss or density loss) by Meentemeyer's graduate student (Smith 1988), and she found that just over 50% of the reported wood decay rates could be explained by AET alone. Wood type explained another 8% (e.g., conifers decayed at 80% the rates of hardwoods) but diameter of the wood pieces, surprisingly, did not contribute to the model. She noted variability in methods as contributing significant sources of variation. Long-term regional studies of wood decay that control many of these methodological sources of variation have been conducted (e.g., Schowalter et al. 1998).

One method that probably has not served wood decay studies very well is the adoption of Olsen's (1963) "k" value to quantify the rate of change in density or mass. This method presumes that decay occurs in a rather orderly fashion and that the same proportion of the material is decayed each year. This technique allowed researchers to measure decay rates in several different ways (e.g., by calculating mass loss over relatively short time periods or using simple estimates of flux into or out of a decaying "pool"). But use of a single exponent has been criticized as being too simple for substrates that were actually mixtures of substance of variable decomposability (Minderman 1968). Calculation of k using estimates of annual inputs divided by estimates of standing stock was noted to produce the highest reported rates of decay (Smith 1988) because standing stocks can be temporarily depressed due to non-steady-state conditions. Another innovation to quickly assess decay rates was the use of chronosequences (a series of sites in which the ages of the decaying wood was known or derived). Errors can arise in assigning ages to logs, and fragmentation from logs can be difficult to estimate. Probably the most accurate way to measure wood decay is to follow fresh logs through time and make repeated measures over the entire course of decay. Few wood decay studies have had this luxury, and none have followed the piece though complete decay. Given that the methods used to quantify wood decay rates can introduce variation, it perhaps should not be so surprising that highly variable rates can come out of different studies.

Despite the interest in the decay rates, it may be that it is more important to understand where the organic matter or carbon (C) goes following decay. Nutrient fluxes from decaying wood have been recently studied (Palviainen et al. 2004; Hafner et al. 2005) as have been C fluxes to the soil (Hafner et al. 2005; Zalamea et al. 2007). The role of wood is being considered in detrital C pools (Harmon and Hua 1991; Liu et al. 2006). Regional C budgeting attempts would be greatly aided if a general factor could be derived for wood that partitions the C flux from wood during decay between that which returns immediately to the atmosphere as CO_2 and that which enters the forest floor and soil and contributes to the "long-term" soil organic matter pools. Kononova (1966) suggested such a factor as ranging from 0.3 to 0.5 for all plant material, but this has never been tested to our knowledge. On a site level, organic matter and nutrient fluxes to the soil during wood decay should increase soil water storage, nutrient availability, and beneficial microbial populations. These, in turn, should benefit plant growth and may be one way of stabilizing the ecosystem during forest regrowth.

Our study of wood decay on WS 7 is different from most studies of wood decomposition in that it follows wood decomposition via repeated samples on individual logs over an 11-year period. This is a sufficiently long period to assess over half the wood mass loss at Coweeta. It includes measures of both solution and gaseous losses during decay. Because it follows measured logs from the start, we are able to make relatively reliable estimates of fragmentation rates. Estimates of fragmentation and solution fluxes allow us to estimate fluxes of organic matter and nutrients to the forest floor.

Site Description

Coweeta Hydrologic Laboratory (lat. 35° 03' N, long. 83° 25' W) is located in the southern Appalachian Mountains in western North Carolina. Coweeta forests are characterized as mixed-aged, oak-hickory hardwoods on steep slopes. Compared to other temperate forests of North America, Coweeta is considered to have a relatively humid (annual precipitation of 180 cm per year, evenly distributed) and moderately warm climate (mean temperature 13°C, with monthly mean temperatures ranging from 2°C to 20°C annually). WS 7 drains a 59-ha basin and was clearcut in 1977. Saw logs were removed and remaining stems, along with slash, were left in place to decompose. Details of the management are given by Swank and Webster (see chapter 1, this volume). Slopes on WS 7 range from 20% to 80%; aspect varies due to the incised valley of the drainage network, but is generally south. The elevation ranges from 772 m at the gage to over 1000 m at the top of the ridge. Soils are Humic Hapludults on the lower elevations and Typic Distrochrepts on the higher elevations. Forest floors are composed of L, F, and H layers with thick H layers forming particularly on sites with *Kalmia latifolia* understory.

Methods

Green volumes of large wood (logs > 5 cm diameter) and densities of selected disks from logs were measured in a series of 4 x 4 m plots during year 1 following clearcutting of the WS 7 forest. Eighty-eight individual logs from 12 different species of trees from the first-year study were relocated in years 6 and 7 (Mattson 1986; Mattson et al. 1987). The logs that could be relocated were tagged and a series of measures made in years 6 and 7. The logs were revisited and measured in year 11 to provide a long-term data set on individual logs as they underwent decomposition.

Selected logs were measured for gaseous fluxes of CO_2 and solution fluxes of C and nutrients. Logs were fitted with 10-cm-diameter chambers attached to the wood surface in which CO_2 efflux was measured monthly via the static method of absorption into NaOH solution (Anderson 1982). Solution flux to the forest floor was estimated by fitting a collection trough beneath entire logs. Leachate volume was periodically measured and subsamples were analyzed for organic C and nutrients. Fragmentation losses were estimated by the assessment of volume losses from the individual logs where volume was measured in year 1. Wood nutrient concentrations were also followed over time. Mass of small wood (diameter < 5 cm) was measured at year 1 and 7 on 2 x 2 m plots. No measures were made at year 11, as all the small wood had decayed by then.

Results and Discussion

The overall density loss from the individual logs generally showed a linear decline; but some logs showed highly variable changes in density over time (figure 7.2). The difference in the slopes of the slowest and fastest logs suggest up to 3-fold differences in decay rates. The differences among species appear to be minor, except the decay-resistant black locust has clearly a slower rate of decay. The individual logs in figure 7.2 are highly variable (increasing and decreasing over time), showing that individual disks cut from the same log can produce highly variable rates of density

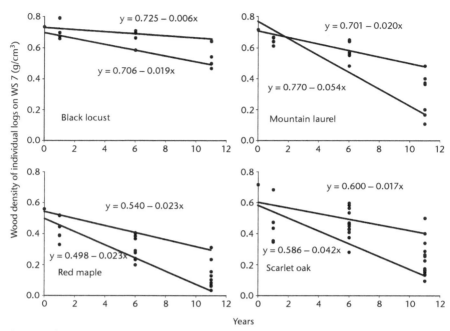

Figure 7.2 Density loss from individual logs and linear regressions of the slowest and the fastest decay shown by individual logs (*solid lines*). These four major species groups show within-species variation.

loss. Schowalter et al. (1998) also observed highly variable rates of loss over time for oak logs, and they suggest a two-phase pattern of rapid decay of inner bark followed by slower decay of sapwood and heartwood. Our data of density, determined on individual disks destructively sampled over time, shows that the density at any single point along a log can be highly variable once decay has started. Our data point out the need to have multiple samples from individual logs in order to assess decay rates. State of decay within an individual log has been shown to be variable spatially (Swift 1977; Graham et al. 1980; Pyle and Brown 1999). Most of the variation in our log decay appeared to be associated only somewhat with species but at least as much with individual conditions of the logs and likely reflect individual wood variation or contact with the soil or even the colonization patterns by decomposers.

The mass of large wood at four time periods (years 0, 1, 6, and 11) was calculated as the product of individual species volume and density for each period. Volume of large wood was measured for each tree species of wood at year 1. This volume was assumed to be equal to year 0 volumes and was corrected for observed volume loss or fragmentation loss by species at each time period (years 6 and 11). Each species volume was multiplied by the mean of its density for that time period. The mass was then summed over all species to get total wood at each time period. Small wood was simply collected and weighed at two time intervals (table 7.1). The

Table 7.1   Wood decomposition on WS 7 over the first 11 years following clearcutting of the forest.

	Amount remaining			
	Year 0	Year 1	Year 6	Year 11
Large wood*				
Green volume (cm³/m²)	17,609	17,609	15,963	10,391
Green density (g/cm³)	0.5176	0.4248	0.3321	0.2021
Dry mass (g/m²)	9,114	7,481	5,302	2,100
Small wood**		Year 1	Year 7	
Dry mass (g/m²)		2,100	780	

* Large wood > 5 cm diameter
**Small wood < 5 cm diameter

interacting patterns of density loss and volume loss, and the resulting decline of the entire large wood mass are shown in figure 7.3. The mass-weighted density of large wood was calculated as the ratio of summed mass divided by total volumes from table 7.1. This mass-weighted density declined in a negative exponential fashion while volume showed no changes and then began to decline at an increased pace. These two different processes contribute to mass loss that appeared to be linear over time as has been suggested for slow decay material such as wood (Taylor and Parkinson 1988).

The relationship of increasing fragmentation occurring as wood density decreased is shown for year 11 data in figure 7.4. Most fragmentation of logs did

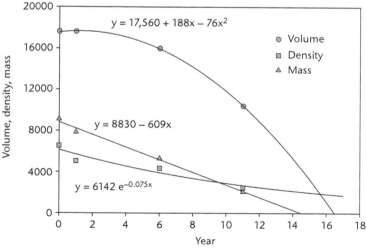

Figure 7.3   Wood decomposition on WS 7 versus time since clearcutting. The processes of density loss and fragmentation are compared to show how each contributes to mass loss. Solid lines are curve fits to the four data points. These are extrapolated to year 17. Density is the mass-weighted mean density of all measured wood pieces. The data have been scaled to allow plotting on the same axis: volume (cm³/m²), mass (g/m²), and density (g/cm³ x 10,000).

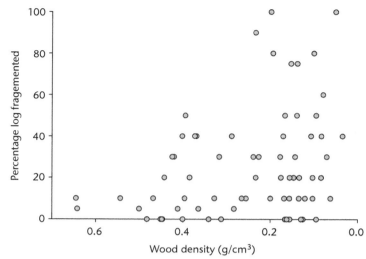

Figure 7.4   Degree of fragmentation of log versus density of wood for wood pieces on WS 7 at year 11. Degree of fragmentation is the percent of the original log that had completed fragmented or disintegrated. The mean wood density of class 3 and class 4 logs, or those that showed signs of complete structural loss, was 0.176 g/cm^3. This was used as an estimate of wood density when the remaining logs would be considered to have entered the forest floor via fragmentation or collapse.

not begin to occur until after wood densities fell below 0.5 g/cm^3 and many pieces underwent complete fragmentation at densities as high as 0.25 g/cm^3. The point at which fragmentation begins to occur is particularly important as it represents the point where the woody material contributes significantly to the forest floor organic matter.

Most wood decay studies do not address the later stages of decay; but in our study, wood was placed on the ground as a single "cohort," where age was known, and because the wood decayed rapidly, we were able to project the complete collapse of a volume of large wood in about 16 years (figure 7.3). Gore and Patterson (1986) assembled wood data from four studies of the clearcutting of hardwood forests of the northeast United States. Their projections match figure 7.3 closely, but their rates of loss were slower; initial wood masses of 10,500 g/m^2 declined to minimums of less than 2,000 by year 30. Afterward, they showed that wood would begin to recover mass as inputs from wood litterfall began to accumulate. They projected steady-state masses of 4,000 g/m^2.

We partitioned the flux of C from wood as it decayed during years 7 through 11. We used our CO$_2$ effluxes, solution fluxes, and fragmentation rates to show that two-thirds of the C losses from wood went to the atmosphere as CO$_2$ and one-third entered the forest floor as either fragmentation or solution fluxes (table 7.2). The CO$_2$ fluxes from wood measured at years 6 and 7 (table 7.2) were 30% higher than the losses of C if we used the overall density losses (table 7.1). Using these data, we show the patterns of C fluxes shift over time as going mostly to the atmosphere

Table 7.2   Annual carbon fluxes associated with wood* on WS 7 estimated for years 7–11.

Flux	kg C ha⁻¹–yr⁻¹
Solution inputs from canopy throughfall	+27
Solution outputs via leaching	−145
Outputs via fragmentation	−1190
Outputs via CO_2	−2715
Net change	−4023

* Wood was assumed to be 50% C.

in the first years, and then going mostly to the forest floor in the later years (figure 7.5). When we integrated the two regression equations in figure 7.5 and solved for the time period between year 0 and year 16.5 (the point of projected complete collapse), we obtained the following total projected fluxes of wood organic matter: 9,700 g/m² of total loss and 4,100 g/m² of flux to the forest floor. This gives a slightly higher partition coefficient of 40% of the total loss of wood transferred to the forest floor. In either case, these fluxes to the forest floor agree with the general humus "coefficient" proposed by Kononova (1966); that is, 0.3 to 0.5 of the mass of fresh plant material contributes to humus formation. Graham et al. (1980) partitioned fragmentation and density loss from dead balsam fir in chronosequences of fir-waves and calculated a partition twice as high as ours. They estimated that two-thirds of their wood fragmented and entered the forest floor and one-third was lost to density decreases.

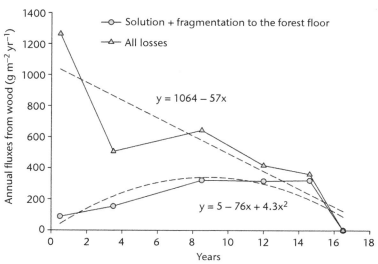

Figure 7.5   Annual fluxes of organic matter from wood during decomposition on WS 7. The two sets of symbols indicate the total losses from wood and those losses that go to the forest floor, respectively. The dashed lines are regressions. The area between the two dashed lines is the loss going to the atmosphere as CO_2; the area under the bottom dashed line is the flux to the forest floor.

Table 7.3   Nutrients in large wood on WS 7 with comparison to other studies.

| | Nutrient concentration (% by weight) | | | Nutrient mass (g/m² of watershed area) | | |
	Year 6	Year 11	Range of published values[1]	Wood Year 6	Wood Year 11	Live boles on WS 18[2]
N	0.186	0.317	0.14–0.30	9.86	6.62	26
P	0.008	0.012	0.002–0.02	0.40	0.25	11
Ca	0.159	0.273	0.02–0.26	8.42	5.70	8
K	0.060	0.038	0.01–0.10	3.19	0.79	14
Mg	0.010	0.100	0.01–0.03	0.53	2.09	1.8
Na	0.001	0.008	0.001	0.04	0.16	–

Note: Concentrations are for samples of large wood tissue (not bark). Mass estimates were extrapolated to both wood and bark (assuming bark concentrations were similar to wood concentrations).
[1] Grier (1978), Graham et al. (1980), Foster and Lang (1982), Barber and Van Lear (1984), Schowalter et al. (1998).
[2] Monk and Day (1988). Live wood mass on WS 18 was 13,400 g/m²; dead wood masses on WS 7 were 5,300 and 2,100 g/m² for years 6 and 11, respectively.

The major nutrients contained in the wood at years 6 and 11 were low, as expected of wood (table 7.3). Compared to other studies, our concentrations were slightly high in nitrogen (N), calcium (Ca), and magnesium (Mg), and low in phosphorus (P) and potassium (K). The concentrations of nutrients generally increased from year 6 to year 11, except for a slight decrease in K concentration. K decrease may reflect greater solubility and its loss may reflect physical leaching. The increase in concentration of the other nutrients, particularly N, indicates at least passive, or more likely active, retention by decomposing fungi. Organic N has been known to increase wood decay rates in laboratory studies (Campbell 1952; Cowling and Merrill 1966), and rates of microbial N fixation have been shown to increase in proportion to the degree of wood decay (Jurgensen et al. 1984). In general, other studies have shown that most nutrients increase in concentration during the course of wood decay, except for K, which has been observed to decrease (Grier 1978; Foster and Lang 1982; Fahey 1983; Keenan et al. 1993; Barber and Van Lear 1984).

Despite the low concentrations of nutrients in the wood, the mass of nutrients was still relatively large because of the large mass of wood. The mass of nutrients in the live boles on WS 18 (table 7.3) show how nutrients are conserved or lost in the wood. The mass of wood was considerably smaller than the mass of live boles on WS 18, but N was somewhat conserved as were Ca and Mg. P and K were considerably reduced, as these did not seem to be conserved in wood.

Estimates of element transfers from wood to the forest floor via leaching were collected in a set of logs placed onto modified throughfall collectors (table 7.4). The solution collected was often a brown color, indicating high content of dissolved organic C. The average concentration of dissolved organic C in year 7 was quite high at 115 mg/L, with the highest concentration of 490 mg/L from the most highly decayed logs. The calculated fluxes of dissolved nutrients (Ca, Mg, K) appeared to be relatively important to the soil directly beneath the logs, but since the projected cross sectional area of the logs equaled only 14% of the surface of the watershed, the fluxes on a watershed level were small. For example, Ca flux in solution to the

Table 7.4   Concentrations (mg/L) of dissolved organic carbon (DOC) and elements in solution passing into and out of wood and the resulting net flux to the forest floor for year 7 on WS 7.

	Canopy throughfall	Wood leachate	Net change in concentration	Flux from wood to forest floor (g m^{-2}yr^{-1})
DOC	5–24	56–180	76	102
N-NO$_3$	0.11	0.08	−0.03	0.0
N-NH$_4$	0.10	0.17	0.07	0.1
P-PO$_4$	0.05	0.07	0.02	0.0
Ca	0.73	3.32	2.59	2.7
K	2.30	3.71	1.41	1.5
Mg	0.26	0.68	0.42	0.4

Note: Fluxes were calculated using the net increase in concentration from throughfall to wood and a measured annual solution throughput of 136 cm through wood. The fluxes represent the transfer to forest floor area directly below the wood, which is 14.25% of the total forest floor area. To get fluxes to the entire watershed, the fluxes must be multiplied by 0.1425.

entire watershed was 0.4 g m^{-2} yr^{-1} and equal to about 7% of the estimated forest uptake rates (5.4 g m^{-2} yr^{-1}) for a reference forest on nearby WS 18 (Monk and Day 1988). The fluxes of nitrate, phosphate, and ammonium were negligible, due to the low concentrations and apparently conservative nature of the decomposing wood for these nutrients. This pattern of nutrient fluxes is similar to that observed for slash measured in year 1 after cutting. Generally, N uptake was observed and base cation leaching was substantial (Knoepp et al., chapter 4, this volume).

The flux of organic C in solution was not a large loss or a large flux to the detrital C pools. For example, the annual flux of dissolved C at year 7 from wood was calculated at 11 gC/m^2 of watershed area. This was less than 1% of the soil C content of the top 10 cm of soil (1,806 gC/m^2, Mattson and Swank 1989) as measured in year 7. Therefore solution fluxes from wood could not readily be the cause for the observed large, but short-lived, increase of concentrations of organic matter in the soil during the first two years following clearcutting of WS 7 (Waide et al. 1988; Knoepp and Swank 1997; see Knoepp et al., chapter 4, this volume). Still, the solution flux of dissolved C and nutrients are likely important as they create small micro-sites of nutrient enrichment. The leachate from wood is not uniformly distributed over the woody material but instead drips from distinct points and areas of contact with the soil.

While the transfers of material via solution from wood to the forest floor were relatively small, the transfers via fragmentation of the wood were notably greater (table 7.5). The total inputs of wood to the forest floor over the entire wood decomposition cycle, estimated to be about 16.5 years, was 4,135 g/m^2, for an average rate of input of 250 g m^{-2} yr^{-1}. This input is 60% of the annual litterfall inputs to the forest floor at year 6 (420 g/m^2; Mattson 1986) and clearly should have affected organic matter pools. Indeed, the forest floor measures on WS 7 showed a trend of increasing mass following clearcutting. Precut forest floor mass (L and F layers only) was 816 g/m^2 versus first-year postcut mass of 970 g/m^2 (Seastedt and Crossley 1981). By year 6, the forest floor (L and F layers) of WS 7 was 1158 g/m^2 versus

Table 7.5 Estimated transfers (g/m²) of dry mass and nutrients to the forest floor from fragmenting wood on WS 7.

	Years 0–6	Years 6–11	Years 11–17	Total
Large wood*	570	1190	1410	3170
Small wood**	1120	780	0	1900
N	3.14	4.95	4.47	12.57
P	0.14	0.20	0.17	0.50
Ca	2.69	4.26	3.85	10.79
K	1.01	0.97	0.54	2.52
Mg	0.17	1.08	1.41	2.66
Na	0.02	0.09	0.11	0.22

* Large wood > 5 cm diameter
** Small wood < 5 cm diameter

equivalent measures on nearby uncut WS 2 of 880 g/m² (Mattson 1986). Increase in the forest floor is, in part, attributed to a temporary reduction in decomposition rates due to extremes of surface temperatures and desiccation during the first few years following the clearcut (Seastedt and Crossley 1981; Abbott and Crossley 1982). The inputs of wood are considered an important and perhaps larger cause of the observed forest floor increases. However during year 20, Vose and Bolstadt (2007) did not detect larger forest floor mass (L, F, and H layers) on WS 7 versus WS 2 based on only three sample sites.

Seven years after clearcutting, soil C was measured in WS 7 and adjacent WS 2 (Mattson and Swank 1989). The 0–10 cm layer contained 1806 g/m², a value close to that on WS 2 (2106 g/m²) and also similar to values 8 years after cutting (1647 g/m²) reported by Knoepp et al. (see chapter 4, this volume). The response of soil cations, C, and N are presented over a 33-year period after cutting WS 7 by Knoepp et al. (chapter 4, this volume).

The flux of nutrients to the forest floor from fragmenting small and large wood was substantial (table 7.5). The combined fragmentation of large and fine wood turns out to be relatively constant input of nutrients as small wood and bark fragmented early and large wood fragmented later. In this calculation, small wood and bark were assumed to have the concentrations shown for large wood (table 7.3). The total flux for the entire decay cycle divided by 16.5 years (the approximate time for wood to completely enter the forest floor) gives average annual flux rates that were equal to about one-quarter the amounts in litterfall on nearby control WS 18. For example, calculated N flux of fragmented wood averaged 0.76 g m⁻² yr⁻¹ compared to reported litterfall N flux of 3.3 g m⁻² yr⁻¹ for control WS 18 (Monk and Day 1988). Average Mg fluxes were 0.16 versus 0.65 g m⁻² yr⁻¹ for litterfall on WS 18.

Synthesis

The decomposition rate of the mass of wood on WS 7 was rapid but similar to that found in other reports from the southeastern United States for hardwood decay

(Abbott and Crossley 1982; Harmon 1982) or pine decay (Barber and Van Lear 1984). Wood mass loss is about twice as fast in the southeast versus the northeast United States (Gore and Patterson 1986) and several-fold faster than rates from the Pacific Northwest (Harmon et al. 1986). The interaction of early density loss and later fragmentation loss produced a generally linear rate of mass loss. Most wood decay studies have not followed individual pieces of wood over time but have instead attempted to derive decay rates from existing pieces of wood by estimating the time the wood had been decaying (i.e., chronosequences) and these studies may have underestimated fragmentation losses. But despite this difference, wood density loss was also high in our study.

The conditions of the wood substrate may be one reason for high rates of wood decay at Coweeta. The pieces of wood were relatively short and small in diameter. Average diameter was 12 cm, and few pieces were more than 25 cm. Lengths were typically 1 to 4 meters, and many pieces were cut in two during repeated sampling. Smaller wood has generally been shown to decay more rapidly as a result of greater proportion of nonresistant sapwood (Aho 1974); to have high surface-to-volume ratio, more readily exposing the inner wood to the elements and to decay organisms; and to have greater diffusion rates (Boddy 1983). Smaller pieces of wood also tend to have greater contact with the soil, which promotes moisture, decay fungi, and insects (Barber and Van Lear 1984). Hardwood species have lower lignin contents than conifers (Cote 1977), and lignin is known to be resistant to decay (Campbell 1952). The class of fungi called "white rots" can degrade lignin and are generally present in hardwood while "brown rots" cannot degrade lignin and are predominant in conifer wood (Rypacek et al. 1986). Finally, excessive wood moisture content has been noted as reducing decay in conifers (Progar et al. 2000). Perhaps a more important factor responsible for high wood-decay rates at Coweeta is the high and frequent rainfall in combination with the warm temperatures (Swift et al. 1988). Higher rates of wood decay are reported in warm moist regions, and the highest rates are from the tropics (Lang and Knight 1979; Smith 1988).

It may be important to clarify that the high fragmentation rate was actually a measure of "loss" of wood, and was not precisely "decomposition." We assumed the wood fragments that fell to the forest floor were equivalent to density loss in contributing to wood decay. This may overestimate decomposition rates of wood because wood pieces that fragment may still be composed of partially sound wood. Therefore, fragmentation may be considered as simply moving wood from logs to the forest floor, where it may persist. On the other hand, the fragments are clearly smaller and of lower density than logs and, once in the forest floor, act more as humus. Regardless of the terminology, it is important to differentiate between the loss of C to the atmosphere and the transfer of organic matter to the forest floor. Our data suggest that up to two-thirds of the mass is lost as CO_2 and that one-third or more enters the forest floor and soil as wood fragments. The transfer of wood to the forest floor contributes to refractory organic matter and nutrients. The forest floor increased in mass on WS 7 by year 6. Much of this increase may have been due to wood fragments. For example, wood fragments comprise up to 30% of forest-floor volume in forests in the Adirondak Mountains of New York State (McFee and Stone 1966). In our study, wood fragmentation was approximately equal to 50% of litterfall during the period from 6 to 11 years after clearcutting. The inputs of wood would also be expected to be incorporated into the soil organic matter pool and contribute to long-term C storage

Wood decomposition is not conceptually difficult to measure, and the measured rate is a good of integrator of a basic ecosystem process. The wood decomposition rate should reflect the tendency of a system to accumulate organic matter. Ecosystems have often been characterized by the accumulation of organic matter on site (Odum 1969; O'Neill et al. 1975; Waide 1988). Organic matter accumulation is in turn the result of the balance between the rates of primary production and decomposition. Coweeta forests and most forests of the southeastern United States do not typically have large accumulations of wood. The co-occurrence of moisture and warm temperatures create favorable conditions for rapid wood decay. In contrast, the accumulations of woody organic matter in the coniferous forests of the Pacific Northwest can exceed 300 Mg/ha where wood decomposition rates are among the lowest reported (e.g., k of 0.005 yr^{-1}; Harmon and Hua 1991). Globally, the rates of primary production appear to be more narrowly constrained by climate than are the rates of organic matter decay (Meentemeyer 1984; Harvey 1989). This would mean that accumulations of organic matter are under greater control by decomposition rates that vary more than the production rates. Decomposition rates have a stronger exponential relationship with temperature than do plant growth rates. Therefore, one may expect that in areas of at least moderate primary production, such as a forest, the decay rate of resistant organic matter will be the primary determinant of the accumulation of organic matter.

Decaying wood acts as a slow-release fertilizer, releasing nutrients bound in the wood as the surrounding forest grows. The nutrient content in the wood on WS 7 was generally conserved and nearly equal to the pools in the forest floor. Also, the flux of N, Ca, and Mg in fragmenting wood was to be about one-quarter that of fluxes in litterfall.

Thus wood contributed to ecosystem resilience through woody debris decomposition and the subsequent flux of both organic matter and nutrients to the forest floor, increasing the nutrient content of detrital pools and supplying nutrients to the regrowing forest. As organic matter in wood decomposed into CO_2 and was lost from the system, the regrowth of new vegetation fixed CO_2 into new plant matter. Regrowth of new vegetation during the first two years of regrowth on WS 7 was estimated to be between 2.1 and 3.7 t/ha of biomass (Boring et al. 1988). This rate certainly increased over the first 11 years and surpassed the rate of organic matter loss from wood (5 t ha^{-1} yr^{-1}) as measured by woody CO_2 loss at year 6.

Literature Cited

Abbott, D. T., and D. A. Crossley, Jr. 1982. Woody litter decomposition following clear-cutting. *Ecology* 63: 35–42.

Aho, P. E. 1974. Decay. Pages Q1–Q17 in Environmental effects of forest residues management in the Pacific Northwest: A state of knowledge compendium. Unitred States Forest Service General Technical Report PNW-24.

Anderson, J. P. E. 1982. Soil respiration. Pages 831–871 in *Methods of Soil Analysis*. A. C. Page, R. H. Miller, and D. R. Keeney, editors. American Society of Agronomy and Soil Science Society of America, Madison, Wisconsin.

Barber, B. L., and D. H. Van Lear. 1984. Weight loss and nutrient dynamics in decomposing woody loblolly pine logging slash. *Soil Science Society of America Journal* 48: 906–910.

Boddy, L. 1983. Carbon dioxide release from decomposing wood: effect of water content and temperature. *Soil Biology Biochemistry* 15: 501–510.

Boring, L. R., W. T. Swank, and C. D. Monk. 1988. Dynamics of early successional forest structure and processes in the Coweeta basin. Pages 161–179 in *Forest Hydrology and Ecology at Coweeta*. W. T. Swank and D.A. Crossley, Jr., editors. Springer-Verlag, New York, New York.

Campbell, W. G. 1952. The biological decomposition of wood. Pages 1061–1116 in *Wood Chemistry*. L. E. Wise and E. C. Jahn, editors. Reinhold Pub. Co., New York, New York.

Cartwright, K. S. G., and W. P. K. Findlay. 1958. Decay of timber and its preservation. Her Majesty's Stationary Office, London.

Cote, W. A. 1977. Wood ultrastructure in relation to chemical composition. *Recent Advances in Phytochemistry* 11: 1–44.

Cowling, E. B., and W. Merrill. 1966. Nitrogen in wood and its role in wood deterioration. *Canadian Journal of Botany* 44: 1539–1554.

Fahey, T. J. 1983. Nutrient dynamics of above ground detritus in lodgepole pine (*Pinus contorta* spp. *latifolia*) ecosystems, southeastern Wyoming. *Ecological Monographs* 53: 51–72.

Foster, J. R., and G. E. Lang. 1982. Decomposition of red spruce and balsam fir boles in the White Mountains of New Hampshire. *Canadian Journal of Forest Research* 12: 617–626.

Gore, J. A., and W. A. Patterson, III. 1986. Mass of downed wood in northern hardwood forests in New Hampshire: potential effects of forest management. *Canadian Journal of Forest Research* 16: 335–339.

Graham, R. L., G. E. Lang, and W. A. Reiners. 1980. Loss of mass and chemical change in decaying boles of a subalpine balsam fir forest. *Ecology* 61: 1460–1473.

Grier, C. C. 1978. A *Tsuga heterophylla–Picea stichensis* ecosystem of coastal Oregon: decomposition and nutrient balances of fallen logs. *Canadian Journal of Forest Research* 8: 198–206.

Hafner, S. D., P. M. Groffman, and M. J. Mitchell. 2005. Leaching of dissolved organic carbon, dissolved organic N, and other solutes from coarse woody debris and litter in a mixed forest in New York State. *Biogeochemistry* 74: 257–282.

Harmon, M. E. 1982. Decomposition of standing dead trees in the southern Appalachian Mountains. *Oecologia* 52: 214–215.

Harmon, M. E., J. F. Franklin, F. J. Swanson, P. Sollins, S. V. Gregory, J. D. Lattin, N. H. Anderson, S. P. Cline, N. G. Aumen, J. R. Sedell, G. W. Lienkaemper, K. Cromack, Jr., and K. W. Cummins. 1986. Ecology of coarse woody debris in temperate ecosystems. *Advances in Ecological Research* 15: 133–302.

Harmon, M. E., and C. Hua. 1991. Coarse woody debris dynamics in two old-growth ecosystems. *BioScience* 41: 604–610.

Harvey, L. D. D. 1989. Effect of model structure on the response of terrestrial biosphere models to CO_2 and temperature increases. *Global Biogeochemical Cycles* 3: 137–153.

Hunt, G. M., and G. A. Garratt. 1938. *Wood Preservation*. MacGraw Hill, New York, New York.

Jurgensen, M. F., M. J. Larson, S. D. Spano, A. E. Harvey, and M. R. Gale. 1984. Nitrogen fixation associated with increased wood decay in Douglas-fir residue. *Forest Science* 30: 1038–1044.

Kaarik, A. A. 1974. Decomposition of wood. Pages 129–174 in *Biology of Plant Litter Decomposition*. Vol. 1. C. H. Dickenson and G. T. F. Pugh, editors. Academic Press, London.

Keenan, R. J., C. E. Prescott, and J. P. Kimmins. 1993. Mass and nutrient content of woody debris and forest floor in western red cedar and western hemlock forests on northern Vancouver Island. *Canadian Journal of Forest Research* 23: 1052–1059.

Knoepp, J. D., and W. T. Swank. 1997. Forest management effects on surface soil carbon and nitrogen. *Soil Science Society of America Journal* 61: 928–935.

Kononova, M. M. 1966. *Soil Organic Matter*. Pergamon Press, Oxford, United Kingdom.

Lang, G. E., and D. H. Knight. 1979. Decay rates for boles of tropical trees in Panama. *Biotropica* 11: 316–317.

Liu, W. H., D. M. Bryant, L. R. Hutyra, S. R. Saleska, E. Hammond-Pyle, D. Curran, and S. C. Wofsy. 2006. Woody debris contribution to the carbon budget of selectively logged and maturing mid-latitude forests. *Oecologia* 148: 108–117.

Mattson, K. G. 1986. Detrital carbon budget following clearcutting hardwood forests in the Southern Appalachians. Dissertation. University of Georgia, Athens.

Mattson, K.G., W. T. Swank, and J. B. Waide. 1987. Decomposition of woody debris following clearcutting in the Southern Appalachians. *Canadian Journal of Forest Research* 17: 712–721.

Mattson, K. G., and W. T. Swank. 1989. Soil and detrital carbon dynamics following forest cutting in the Southern Appalachians. *Biology and Fertility of Soils* 7: 247–253.

McFee, W. W., and E. L. Stone. 1966. The persistence of decaying wood in the humus layers of northern forests. *Soil Science Society of America Proceedings* 30: 513–516.

Minderman, G. 1968. Addition, decomposition, and accumulation of organic matter in forest. *Journal of Ecology* 56: 355–362.

Meentemeyer, V. 1984. The geography of organic decomposition rates. *Annals of the Association of American Geographers* 74: 551–560.

Monk, C. D, and F. P. Day, Jr. 1988. Biomass, primary production, and selected nutrient budgets for an undisturbed watershed. Pages 151–159 in *Forest Hydrology and Ecology at Coweeta*. W. T. Swank and D.A. Crossley, Jr., editors. Springer-Verlag, New York, New York.

Odum, E. P. 1969. The strategy of ecosystem development. *Science* 126: 262–270.

Olsen, J. S. 1963. Energy storage and the balance of producers and decomposers in ecological systems. *Ecology* 44: 322–331.

O'Neill, R. V., W. F. Harris, B. S. Ausmus, and D. E. Reichle. 1975. A theoretical basis for ecosystem analysis with particular reference to element cycling. Pages 28–40 in *Mineral Cycling in Southeastern Ecosystems*. F. G. Howell, J. B. Gentry, and M. H. Smith, editors. ERDA Conf-740513. National Technical Information Service, Springfield, Virginia.

Palviainen, M., L. Finér, A. M. Kurka, H. Mannerkoski, S. Piirainen, and M. Starr. 2004. Decomposition and nutrient release from logging residues after clear-cutting of mixed boreal forest. *Plant and Soil* 263: 53–67.

Progar, R. A., T. D. Schowalter, C. M. Freitag, and J. J. Morrell. 2000. Respiration from coarse woody debris as affected by moisture and saprotroph functional diversity in Western Oregon. *Oecologia* 124: 426–431.

Pyle, C., and M. M. Brown 1999. Heterogeneity of wood decay classes within hardwood logs. *Forest Ecology and Management* 114: 253–259.

Rypacek, V., J. Dobry, and A. Dziurzynski. 1986. Wood attack by fungi: dynamics of wood decomposition. Academiae Scientiarum Bohemoslovacae Brno, Prague, Czeck Republic.

Schowalter, T. D., Y. L. Zhang, and T. E. Sabin. 1998. Decomposition and nutrient dynamics of oak *Quercus* spp. logs after five years of decomposition. *Ecography* 21: 3–10.

Seastedt, T. R., and D. A. Crossley, Jr. 1981. Microarthropod response following cable logging and clear-cutting in the southern Appalachians. *Ecology* 62: 126–135.

Smith, D. D. 1988. The effect of climate and substrate quality on the broad scale geography of wood decomposition rates. Master's thesis. University of Georgia, Athens.

Spaulding, P., and J. R. Hansbrough. 1944. Decay of logging slash in the northeast. US Department of Agriculture Technical Bulletin No. 876.

Swift, L. W, Jr., G. B. Cunningham, and J. E. Douglas. 1988. Climatology and hydrology. Pages 35–55 in *Forest Hydrology and Ecology at Coweeta*. W. T. Swank and D.A. Crossley, Jr., editors. Springer-Verlag, New York, New York.

Swift, M. J. 1977. The ecology of wood decomposition. *Science Progress* 64: 179–203.

Taylor, B. R., and D. Parkinson. 1988. Aspen and pine leaf litter decomposition in laboratory microcosms. I. Linear versus exponential models of decay. *Canadian Journal of Botany* 66: 1960–1965.

Vose, J. A., and P. V. Bolstad. 2007. Biotic and abiotic factors regulating forest floor CO_2 flux across a range of forest age classes in the southern Appalachians. *Pedobiologia* 50: 577–587.

Waide, J. B. 1988. Forest ecosystem stability: Revision of the resistance-resilience model in relation to observable macroscopic properties of ecosystems. Pages 383–406 in *Forest Hydrology and Ecology at Coweeta*. W.T. Swank and D.A. Crossley, Jr., editors. Springer-Verlag, New York, New York.

Waide, J. B., W. H. Caskey, R. L. Todd, and L. R. Boring. 1988. Changes in soil nitrogen pools and transformations following forest clearcutting. Pages 221–232 in *Forest Hydrology and Ecology at Coweeta*. W. T. Swank and D.A. Crossley, Jr., editors. Springer-Verlag, New York, New York.

Zalamea, M., G. González, C. Ping, and G. Michaelson. 2007. Soil organic matter dynamics under decaying wood in a subtropical wet forest: effect of tree species and decay stage. *Plant and Soil* 296: 173–185.

8

Recovery of Decomposition and Soil Microarthropod Communities in a Clearcut Watershed in the Southern Appalachians

Liam Heneghan*
Alissa Salmore

Introduction

The recovery of ecosystems after disturbance remains a productive theme for ecological research (Holling 1973; DeAngelis 1992; Pimm 1984; Lindenmayer et al. 2004). Numerous studies have focused either on the reestablishment of biological communities (e.g., Elliott and Swank 1994; Niemela 1997; Kotanen 2004) or on the recovery of ecosystem processes after perturbations (Covington 1981; Mann et al. 1988; Elliott and Swank 1994; Likens and Bormann 1995; Knoepp and Swank 1997; Qualls et al. 2000; Elliott et al. 2002). In the case of decomposer organisms and the processes of organic matter decay and the mineralization of nutrients, the recovery of the biota and ecosystem process can be usefully considered together. To what degree must the organismal community recover before associated processes are restored to predisturbance levels? The question is an important one, especially in the context of ecosystem restoration, where emphasis is often placed on measuring the elements of ecosystem structure (expressed often as mean species composition) or the levels or rates of a key ecosystem function, but rarely on both simultaneously (Ehrenfeld and Toth 1997). Restoration ecology, in practice, has largely been a botanical science (Young 2000), and success is often measured by an evaluation of the recovery of a target plant community over one or two subsequent seasons; data investigating the long-term community viability is rarely documented. A clearer understanding of the link between the community and ecosystem processes needed for resilient communities may lead to and justify greater attention in restoration

* Corresponding author: DePaul University, Environmental Science Program, 1110 West Belden Avenue, Chicago, Illinois 60614-3207 USA

practice to a larger suite of variables. One model for this relationship can be found in soil microarthropods and leaf-litter decomposition processes.

Factors determining the rate of leaf-litter decomposition are both biotic (microbes and decomposer fauna) and abiotic (climate, edaphic factors, substrate resource quality). Both biotic and physical factors prevail in all terrestrial ecosystems; however, their relative importance may vary along environmental gradients (Lavelle et al. 1993). A large physical disruption on the scale of a watershed, such as the clearcutting of a stand, will have a dramatic impact on all these factors, including changes arising from soil compaction (Corns and Maynard 1998; Startsev et al. 1998; Lindo and Visser 2004). Clearcutting is likely to have a long-lasting impact since the disturbance will impose modifications to soil structure, changes in site hydrology, altered litter input, and solar radiation flux changes as succession proceeds, which will continue to influence decomposition factors beyond the duration of the actual disruption event. Both direct and indirect impacts on soil biota can be expected after a clearcut. Direct effects result from the physical disruption of the soil. Indirect effects stem from modification of the abiotic habitat factors regulating communities in the soil. In turn, modified soil communities affect rates of decomposition and mineralization of key nutrients.

Soil fauna have been shown to have an important regulatory role in a variety of ecosystem processes including decomposition and nutrient dynamics (Seastedt 1984; Verhoef 1996; Heneghan et al. 1999; Johnston and Crossley 2002). Microarthropods, mainly free-living mites, Collembola, and Protura, are relatively small contributors to ecological energetics, but they appear to have a disproportionately prominent role in determining soil fertility through their regulatory effect on soil microbial communities (Lussenhop 1992). Although they are by no means the only fauna of importance in the process of decomposition, microarthropods have been implicated as playing a major role in the decomposition of dead organic matter because of their ability to regulate belowground food webs (Seastedt 1984; Lussenhop 1992). Reduction of microarthropod populations can reduce rates of decomposition and can result in modified patterns in nutrient leaching in forest soil (Heneghan and Bolger 1996). Some studies on the recovery of decomposition in clearcut watersheds in northern hardwoods have demonstrated, or at least implied, that decomposition rates accelerate after the disturbance (Covington 1981; Aber et al. 1978). In contrast, studies on the decomposition of leaf litter and wood in many other temperate locations, including in clearcut watersheds at Coweeta Hydrological Laboratory, have indicated that decomposition rates and/ or CO_2 efflux were initially lower than in an adjacent undisturbed control watershed (Abbott and Crossley 1982; Seastedt 1979; Blair and Crossley 1988; Gemesi et al. 1995; Pumpanen et al. 2004; Zerva and Mencuccini 2005). Blair and Crossley (1988) established that decay rates remained depressed 8 years after the disturbance by as much as 46% for *Cornus florida*, the most rapidly decomposing leaf species they examined. All three leaf species they employed in the study decomposed more slowly in the disturbed watershed than in the control. Investigations of the response of decomposer organisms and microbial processes to tree harvesting have been similarly equivocal. Houston et al. (1998) showed very slight microbial responses to clearcutting in two Ontario hardwood forests. Striegl et al. (1998)

reported a reduction in soil respiration rates (measured as CO_2 evolved) as a consequence to clearcutting in Saskatchewan. Siira-Pietikäinen et al. (2001) concluded that forest harvesting had little impact on decomposers at a variety of trophic levels. These studies suggest that no generalizations can be made concerning the impact of clearcutting (Whitford et al. 1981; Will et al. 1983; Binkley 1984).

Studies on decomposition in WS 2 and WS 7 at Coweeta concentrated on tracking changes in this process subsequent to the clearcut in 1977. An assumption was made that microarthropod abundance was similar on WS 2 and WS 7 at Coweeta before WS 7 was cable logged in 1977. After a year, abundance was reduced by more than 50% in the clearcut watershed compared with the control (Abbott et al. 1980; Seastedt and Crossley 1981). Blair and Crossley (1988) suggested that microarthropod abundance, which remained 28% lower in the clearcut area than in the control, may have been responsible for decreased decomposition rates.

In this present study, which was made two decades after the clearcut, we reinvestigated the decomposition dynamics of the litter types examined by Blair and Crossley (1988).

Methods

Litterbags with an internal dimension of 10 x 10 cm, constructed with fiberglass window screen, were used in this study. Recently fallen leaves of chestnut oak (*Quercus prinus* L.), red maple (*Acer rubrum* L.), and dogwood (*Cornus florida* L.) were collected from a number of low-elevation watersheds, pooled by species, and mixed thoroughly (to ensure that all the leaves in a single bag were not derived from a single tree). Approximately 2.5 g of air-dried material was placed in litterbags. Five bags were oven-dried at 95°C to establish relationship between air-dried and oven-dried mass loss. Litterbags of each species were placed in three plots in each of both WS 7 and WS 2 in January 1998. Plots matched for altitude, aspect, and slope. Five bags were taken up immediately to establish mass loss from handling. Litterbags were collected every two weeks for 18 months. Bags were oven-dried at 50°C, the litter reweighed and ground, and subsamples were placed in a muffle furnace at 500°C for four hours to obtain an estimate of ash content. This value was subtracted from oven-dried mass to obtain ash-free mass. Nitrogen and carbon percent of the litter at the beginning of the study were determined by combustion using a Carlo Erba C/N analyzer (instrument NA1500).

Decomposition rates were determined using the single negative exponential decay model (Olson 1963). Significant differences in mass remaining were determined using paired one-tailed t-tests on the decomposition rates determined from the model. These tests assessed the null hypothesis that decomposition rate remained lower in the clearcut area than in the paired control watershed. Using paired t-tests, the first-year decomposition rate of each substrate was compared among years, contrasting results of this study with the study conducted by Blair and Crossley (1988). All statistics were performed on SAS.

Six samples of litter measuring 27 cm by 27 cm were collected from randomly chosen points in each of the two watersheds in July 1998. The litter samples primarily consisted of leaf material, but no attempt was made to separate out coarse woody debris. Prior studies (Heneghan, unpublished) had demonstrated that there were no differences in the volume of litter when the two watersheds were compared. These observations were based upon three replicates from quadrats measuring a meter squared. Microarthropods were extracted from the samples using Tullgren funnels, and the fauna were stored in 70% ethanol before being separated into Prostigmata, Mesostigmata, Oribatei, Collembola, and Protura. Oribatei were sorted to morphospecies. Total abundance of each taxonomic group was recorded and species richness, Shannon diversity, and Shannon evenness were measured for Oribatei. Assemblages of oribatids in each of the samples from both watersheds were analyzed using principal components analysis using Multivariate Statistical Package (MVSP). Similarities among these samples were analyzed using Jaccard's index (Magurran 1988). Differences between proportions of microarthropods were tested using t-tests (arcsine transformed). Potential differences between measures of richness and diversity were also examined using t-tests, after a test for homogeneity of variances.

Results

Twenty-one years after the clearcut, the decomposition rates of all three leaf species were either the same, or in the case of *C. florida* greater than, that of the control watershed (figure 8.1). Decay rates were lower a year after the disturbance and remained lower than in the control watershed 8 years afterward by as much as 46% for *C. florida* (Blair and Crossley 1988) (table 8.1). A comparison of the decomposition rates of each leaf species 8 years and 21 years after the clearcut revealed some differences between these years in decomposition rates in the clearcut watershed and the same leaf species in the control watershed. Differences in decomposition rates of chestnut oak in the cleared watershed were marginally significant between years ($t = 2.26$, $P = 0.087$). No differences were detected in the control watershed. Differences for *C. florida* were significant when our results were compared with Blair and Crossley's (1988) data from the clearcut watershed ($P = 0.0004$) and the control watershed ($t = 11.34$, d.f 4, $P = 0.003$). Differences were detected between maple decomposition rates between years in the clearcut watershed ($P = 0.03$) and the control watershed ($P = 0.004$).

Microarthropod Inventory

Twenty-one years after the clearcut watershed, total microarthropod density was greater in WS 7 than in the control watershed (table 8.2). Among the constituent groups of microarthropods, significant differences were detected only for oribatid mites (table 8.2). Oribatid mites were the most prevalent microarthropods, representing almost 50% of the fauna in WS 2 and approximately 57% in WS 7. There were no significant differences detected in the proportion of animals in any of the microarthropod groups (table 8.3). Jaccard's similarity index shows a consistent similarity in oribatid assemblage between samples drawn from the same watershed (figure 8.2).

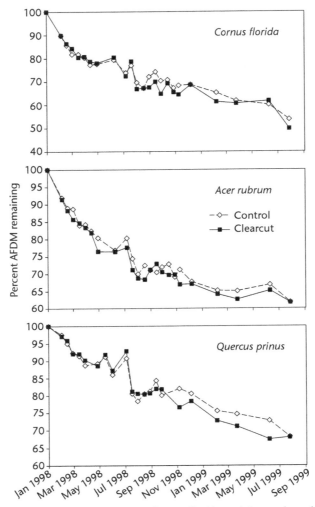

Figure 8.1 Mass loss of *Quercus prinus, Cornus florida,* and *Acer rubrum* in clearcut and control watershed at Coweeta Hydrologic Laboratory, North Carolina.

Oribatid diversity in the two watersheds differed (figure 8.3). There was elevated species richness of oribatid mites in the clear-cut watershed ($P = 0.04$). Differences in Shannon diversity were marginally significant ($t = 2.33$, d.f. = 6.3, $P = 0.056$) (table 8.4). No differences were found in Shannon evenness.

Discussion

Twenty-one years after the clearcut, decomposition rates in WS 7 had reconverged with or surpassed (in the case of one substrate, *C. florida*) the decomposition rates

Table 8.1 Decomposition rates (y^{-1}), percent original mass remaining after 12 months in parentheses.

	Precut		Year 1 (1978)	
	Clearcut (1974-75)	Control (1975-76)	Clearcut	Control
Quercus prinus	0.37 (69.3)	0.29 (72.2)	0.29 (76.7)	0.43 (65.2)
Cornus florida	1.31 (27.8)	0.71 (47.8)	0.63 (55.6)	1.17 (34.3)
Acer rubrum	0.53 (49.0)	0.48 (57.9)	0.37 (70.5)	0.62 (53.7)
	Year 8 (1985)		Year 21 (1998)	
	Clearcut	Control	Clearcut	Control
Quercus prinus	0.19 (78.1)	0.27 (73.1)	0.28 (72.73)	0.24 (75.54)
Cornus florida	0.71 (51.3)	0.85 (45.8)	0.38 (61.32)	0.32 (64.98)
Acer rubrum	0.55 (62.4)	0.64 (52.9)	0.34 (65.22)	0.35 (64.16)

Sources: Precut and year 1 data, Abbott and Crossley (1982); year 8 data, Blair and Crossley (1988).

Table 8.2 Mean and standard error of abundance of each microarthropod group from six samples from both clearcut (WS 7) and control (WS 2) watersheds.

	WS 7 Abundance (Mean, S.E.)	WS 2 Abundance (Mean, S.E.)	Significance
Collembola	52, 12.04	28, 6.64	ns
Astigmata	119, 38.24	66, 27	ns
Megostigmata	27, 8.23	16, 7.96	ns
Oribatida	302, 44.73	134, 47	*
Protura	28, 26.49	6, 4.66	ns
Prostigmata			ns
Others	59, 22.09	19, 14.21	ns
Total	530, 70.68	260, 85.91	*

* = difference at $P < 0.05$ level

Table 8.3 Proportion of abundance in each microarthropod group from six samples from both clearcut (WS 7) and control (WS 2) watersheds.

	WS 7	WS 2
Astigmata	0.03%	0.13%
Collembola	9.74%	9.96%
Mesostigmata	5.25%	5.44%
Oribatida	56.90%	55.87%
Protura	5.37%	4.76%
Prostigmata	22.40%	23.03%
Others	11.06%	10.34%
Total	100.00%	100.00%

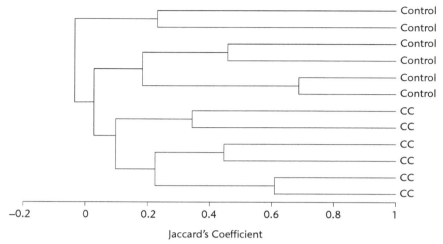

Figure 8.2 Dendrogram of Jaccard's coefficient for oribatid morphospecies from control (WS 2) and clearcut (WS 7) watersheds.

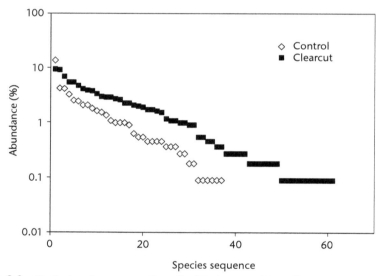

Figure 8.3 Rank abundance curves for oribatid mite assemblages from control (WS 2) and clearcut (WS 7) watersheds.

Table 8.4 Species richness and Shannon diversity and evenness, with associated test statistic for clearcut and control watersheds.

	Clearcut (WS 7) (mean, s.e.)	Control (WS 2) (mean, s.e.)	Significance
Species richness	34, 2.64	21.66, 4.43	*
Shannon diversity	1.31, 0.04	1.05, 0.10	n.s
Shannon evenness	0.86, 0.01	0.84, 0.04	n.s.

* = difference at $P < 0.05$ level

Table 8.5 Percent nitrogen and carbon: nitrogen mass ratios for three substrates used in 1984 and 1996 experiment.

		%N	C:N
1984	Oak	0.87	55.7
1996	Oak	0.88	55.0
1984	Dogwood	0.56	85.7
1996	Dogwood	0.96	50.7
1984	Maple	0.75	62.9
1996	Maple	0.98	47.5

of an adjacent control watershed. Previous observations up to year 8 showed slower decomposition rates in the disturbed watershed compared with the control.

Though decomposition in the treatment and control rates were similar, the rates after 21 years were slower than they were after 8 years. That is, both *C. florida* and *A. rubrum* decomposed more slowly in both the clearcut and control watersheds. *Q. prinus* decomposition rates remained the same in the control watershed but slowed marginally in WS 7. The percent N in the litter of both of these litter types was greater in year 21 (table 8.5). Percent N in leaves can vary greatly from tree to tree and from year to year for different species (Nordell and Karlsson 1995). Although there is often an increase in decomposition rates associated with elevated leaf nitrogen content, analysis of percent N does not reliably indicate substrate quality, as this nitrogen can be incorporated into a variety of recalcitrant molecules (Taylor et al. 1989). Arguably, differences in climate between the two years that we compared could be partially responsible for the differences in decomposition rates. To test this we used monthly climatic data collected from the National Climatic Data Center and made available through the Coweeta Long-Term Ecological Research (LTER) program webpage (http://coweeta.ecology.uga.edu/). The average temperature was warmer by about 1°C for year 21 than for year 8; but total precipitation in both summer and fall was lower in year 21 than in year 8. Lower moisture conditions have important implications for decomposition rates and may have been responsible for the observations of lower decomposition rates for *C. florida* and *A. rubrum*.

These results suggest that in the clearcut watershed there has been a trend toward a return to predisturbance levels of a variety of factors, both abiotic and biotic, that ultimately control the breakdown of leaf material. Blair and Crossley (1988) attributed the substantially lower decomposition rates after 8 years in the clearcut watershed compared to the control to differences in microarthropod abundance. In their study, they noted that microclimatic extremes were greater in the clearcut watershed than in the control. This, they argued, created a situation unfavorable for microarthropods. Though we have not presented data on microclimate here, the large abundances of microarthropods in the disturbed watershed suggest that the abiotic constraints on the growth rates of faunal populations that initially occurred after the clearcut have abated.

Prior to clearcut it is assumed that there were no significant differences between microarthropod densities in the two watersheds (Abbott et al. 1980; Seastedt and Crossley 1981). In the study by Blair and Crossley (1988), the populations were shown to have recovered somewhat from the strong reductions noted after year 1 (remaining 28% lower in the clearcut watershed in the Blair and Crossley study). There was a differential response among the microarthropod groups. Mesostigmata and oribatid mites in the clearcut site were still 50% less abundant compared to the control. Abbott et al. (1980) found that within a year of the cut, the proportional abundance of the most dominant species, *Tectocepheus velatus*, was only 14.5%. The proportional abundance of the most dominant species in the control watershed, *Oppiella nova*, was 27.1%.

A comparison of microarthropod abundance in the two watersheds reveals a greater density of microarthropods in WS 7. The elevated density of oribatid mites in WS 7 was largely responsible for this overall increase. The proportions of organisms in each broad taxonomic group were similar. The proportion of oribatids was the highest of all groups examined, and this contrasts with results from 8 years after the clearcut when prostigmatid mites were proportionately the most prevalent. Blair and Crossley (1988) reported on animals extracted directly from the litterbags, whereas we looked at extractions from the forest floor adjacent to the litterbags. There is no reason to suspect that litterbags should be colonized preferentially by different microarthropod taxonomic groups (Heneghan et al. 2002). Differences in diversity were detected and the Jaccard's similarity measure suggested that samples drawn from each watershed are self-similar.

We have suggested that studies on recovery of ecosystems from large disturbances may be instructive for restoration practice. Ecosystem successional development at WS 7 was not assisted by restoration management. The changes over time in WS 7 reported for the vegetation (see Boring et al., chapter 2, this volume), soil nutrient dynamics (see Knoepp et al., chapter 4, this volume), and hydrology (see Swank et al., chapter 3, this volume) are those that occur in the absence of deliberate human management. Although different processes may proceed at different paces, it is clear that these are mechanistically related. In the data we present here, decomposition rates and the biota that contribute to the decomposition process are linked. The recovery of the former is dependent to some extent on the recovery of the latter. It is also clear that the decomposer system is influenced by the successional patterns of the vegetation; and furthermore, that decomposition may influence primary productivity and vegetation patterns. In contrast to the postdisturbance events that we report on here, restoration management represents, by definition, an attempt to assist in the recovery of an ecosystem that has been degraded, damaged, or destroyed (SER 2002). However, since many (though by no means all) restoration projects focus on manipulating the plant community, and evaluate outcomes based on plant populations or community measures, opportunities for improving outcomes may be neglected if other available indicators, such as microarthropods and decomposition rates, are not considered. Our results suggest that, even in the absence of management, the decomposer subsystem in a hardwood forest re-established itself after two decades. Even if the plant communities differ after two decades, the recovery of decomposition will nevertheless be important in influencing vegetation development. Incorporating soils ecological knowledge into restoration may improve restoration outcomes (Callaham et al. 2008; Heneghan et al. 2008). We suggest two avenues of research, exploring the questions: What is the impact of specific restoration practices (including trampling by project personnel and heavy equipment, etc.) on soil abiotic regimes, biota, and subsequent community development? Are restoration strategies that aim to rehabilitate soil processes more likely to produce resilient ecological results as measured, for instance, by resistance to reinvasion by exotic species, the removal of which is often a prelude to restoration practice)?

Acknowledgments

This research was supported by NSF grant 9707461. Dr. John Blair provided data from his prior work on this system, for which we are very grateful. We would like to acknowledge the encouragement and support of Dr. Wayne Swank in executing this study.

Literature Cited

Abbott, D. T., T. R. Seastedt, and D. A. Crossley, Jr. 1980. The abundance, distribution and effects of clear cutting on oribatid mites (Acari: Cryptostigmata) in the southern Appalachians. *Environmental Entomology* 9: 618–623.

Abbott, D. T., and D. A. Crossley, Jr. 1982. Woody litter decomposition following clear cutting. *Ecology* 63: 35–42

Aber, J. D., D. B. Botkin, and J. M. Melillo. 1978. Predicting the effects of different harvesting regimes on forest floor dynamics in northern hardwoods. *Canadian Journal of Forest Research* 8: 306–315.

Binkley, D., 1984. Does forest removal increase rates of decomposition and nitrogen release? *Forest Ecology and Management* 8: 229–233.

Blair, J. M., and D. A. Crossley, Jr. 1988. Litter decomposition, nitrogen dynamics and litter microarthropods in a southern Appalachian hardwood forest 8 years following clearcutting. *Journal of Applied Ecology* 25: 683–698.

Callaham, M. A., C. C. Rhoades, and L. Heneghan. 2008. A striking profile: soil ecological knowledge in restoration management and science. *Restoration Ecology* 16: 604–607.

Corns, I. G. W., and D. G. Maynard. 1998. Effects of soil compaction and chipped aspen residue on aspen regeneration and soil nutrients. *Canadian Journal of Soil Science* 78: 85–92.

Covington, W. W. 1981. Changes in forest floor organic matter and nutrient content following clearcutting in northern hardwoods. *Ecology* 62: 41–48.

DeAngelis, D. L. 1992. *Dynamics of Nutrient Cycling and Food Webs*. Chapman and Hall, London, United Kingdom.

Ehrenfeld, J. G., and L. Toth. 1997. Restoration ecology and the ecosystem perspective. *Restoration Ecology* 5: 307–318.

Elliott, K. J., and W. T. Swank. 1994. Changes in tree species diversity after successive clearcuts In the southern Appalachians. *Vegetatio* 115: 11–18.

Elliott, K. J., L. R. Boring, and W. T. Swank. 2002. Aboveground biomass and nutrient accumulation 20 years after clear-cutting a southern Appalachian watershed. *Canadian Journal of Forest Research* 32: 667–683.

Gemesi, O., D. Skambracks, and W. Topp. 1995. Effects of clear-cutting on litter decomposition and the density of earthworms in mountain forests of southern Germany. *Forstwissenschaftliches Centralblatt* 114: 272–281.

Heneghan, L., and T. Bolger. 1996. Effects of the components of "acid rain" on the contribution of soil microarthropods to ecosystem function. *Journal of Applied Ecology* 33: 1329–1344

Heneghan, L., C. Clay, and C. Brundage. 2002. Observations on the initial decomposition rates and faunal colonization of native and exotic plant species in a urban forest fragment. *Ecological Restoration* 20: 108–111.

Heneghan, L., D. C. Coleman, X. Zou, D. A. Crossley, Jr., and B. L. Haines. 1999. Soil microarthropod contributions to decomposition dynamics: Tropical-temperate comparisons of a single substrate. *Ecology* 80: 1873–1882.

Heneghan, L., S. P. Miller, S. Baer, M. A. Callaham, J. Montgomery, M. Pavao-Zuckerman, C. C. Rhoades, and S. Richardson. 2008. Integrating soil ecological knowledge into restoration management. *Restoration Ecology* 16: 608–617.

Holling, C. S. 1973. Resilience and stability of ecological systems. *Annual Review of Ecology and Systematics* 4: 1–23.

Houston, A.P. C., S. Visser, and R. A. Lautenschlager. 1998. Microbial processes and fungal community structure in soils from clear-cut and unharvested areas of two mixed wood forests *Canadian Journal of Botany* 76: 630–640.

Johnston, J. M., and D. A. Crossley, Jr. 2002. Forest ecosystem recovery in the southeast US: soil ecology as an essential component of ecosystem management. *Forest Ecology and Management* 155: 187–203.

Knoepp, J. D., and W. T. Swank. 1997. Long-term effects of commercial sawlog harvest on soil cation concentrations. *Forest Ecology and Management* 93: 1–7.

Kotanen, P. M. 2004. Revegetation following soil disturbance and invasion in Californian meadow: a 10-year history of recovery. *Biological Invasions* 6: 245–254.

Lavelle, P., E. Blanchart, A. Martin, S. Martin, A. Spain, F. Toutain, I. Barois, and R. Schaefer. 1993. A hierarchical model for decomposition in terrestrial ecosystems: application to soils of the humid tropics. *Biotropica* 25: 130–150.

Likens, G. E., and F. H. Bormann. 1995. *Biogeochemistry of a forested ecosystem.* Springer-Verlag, New York, New York.

Lindenmayer, D. B., D. R. Foster, J. F. Franklin, M. L. Hunter, R. F. Noss, F. A. Schmiegelow, and D. Perry. 2004. Salvage harvesting policies after natural disturbance. *Science* 303: 1303.

Lindo, Z., and S. Visser. 2004. Forest floor microarthropod abundance and oribatid mite (Acari: Oribatida) composition following partial and clear-cut harvesting in the mixed wood boreal forest. *Canadian Journal of Forest Research* 34: 998–1006.

Lussenhop, J. 1992. Mechanisms of microarthropod microbial interactions in soil. *Advances in Ecological Research* 23: 1–33.

Magurran, A. E. 1988. *Ecological Diversity and Its Measurement.* Chapman and Hall, London.

Mann, L. K., D. W. Johnson, D. C. West, D. W. Cole, J. W. Hornbeck, C. W. Martin, H. Riekerk, C. T. Smith, W. T. Swank, L. M. Tritton, and D. H. Van Lear. 1988. Effects of whole-tree and stem-only clearcutting on postharvest hydrologic losses, nutrient capital, and regrowth. *Forest Science* 34: 412–428.

Niemela, J. 1997. Invertebrates and boreal forest management. *Conservation Biology* 11: 601–610.

Nordell, K. O., and P. S. Karlsson. 1995. Resorption of nitrogen and dry matter prior to leaf abscission: variation among individuals, sites and years in the mountain birch. *Functional Ecology* 9: 326–333.

Olson, J. S. 1963. Energy storage and the balance of producers and consumers in ecological systems. *Ecology* 44: 322–331.

Pimm, S. 1984. The complexity and stability of ecosystems. *Nature* 307: 321–326.

Pumpanen, J., C. J. Westman, and H. Ilvesniemi. 2004. Soil CO_2 efflux from a podzolic forest soil before and after forest clear-cutting and site preparation. *Boreal Environment Research* 9: 199–212.

Qualls, R. G., B. L. Haines, W. T. Swank, and S. W. Tyler. 2000. Soluble organic and inorganic nutrient fluxes in clearcut and mature deciduous forests. *Soil Science Society of America Journal* 640: 1068–1077.

Seastedt, T. M. 1979. Microarthropod response to clear-cutting in the southern Appalachians: effects on decomposition and mineralization of litter. PhD dissertation. University of Georgia, Athens.

Seastedt, T. M. 1984. The role of microarthropods in decomposition and mineralization processes. *Annual Review of Entomology* 29: 25–46.

Seastedt, T. M., and D. A. Crossley, Jr. 1981. Microarthropod response following cable logging and clearcutting in the southern Appalachians. *Ecology* 62: 126–135.

SER (Society for Ecological Restoration Science & Policy Working Group). 2002. The SER primer on ecological restoration. Available at www.ser.org.

Siira-Pietikäinen, A., J. Pietikäinen, H. Fritze, and J. Haimi. 2001. Short-term responses of soil decomposer communities to forest management: clear felling versus alternative forest harvesting methods. *Canadian Journal of Forest Research* 31: 88–99.

Startsev, N. A., D. H. McNabb, and A. D. Startsev. 1998. Soil biological activity in recent clearcuts in west-central Alberta. *Canadian Journal of Soil Science* 78: 69–76.

Striegl, R., G. Kimberly, and P. Wickland. 1998. Effects of a clear-cut harvest on soil respiration in a jack pine–lichen woodland. *Canadian Journal of Forest Research* 28: 534–539.

Taylor, B. R., D. Parkinson, W. F. J. Parsons. 1989. Nitrogen and lignin content as predictors of litter decay rates: a microcosm test. *Ecology* 70: 97–104.

Verhoef, H. A. 1996. The role of soil microcosms in the study of ecosystem processes. *Ecology* 77: 685–690.

Whitford, W. G., V. Meentemeyer, T. R. Seastedt, K. Cromack, Jr., D. A. Crossley, Jr., P. Santos, R. L. Todd, and J. B. Waide. 1981. Exceptions to the AET model: deserts and clear-cut forests. *Ecology* 62: 275–277.

Will, G. M., P. D. Hodgkiss, and H. A. I. Madgwick. 1983. Nutrient loss from litterbags containing *Pinus radiata* litter: influence of thinning, clearfelling, and urea fertilizer. *New Zealand Journal of Forest Science* 13: 291–304.

Young, T. P. 2000. Restoration ecology and conservation biology. *Biological Conservation* 92: 73–83.

Zerva, A., and M. Mencuccini. 2005. Short-term effects of clearfelling on soil CO_2, CH_4, and N_2O fluxes in a Sitka spruce plantation. *Soil Biology & Biochemistry* 37: 2025–2036.

9

Watershed Clearcutting and Canopy Arthropods

Barbara C. Reynolds*
Timothy D. Schowalter
D. A. Crossley, Jr.

Introduction

The southern Appalachian forests are home to myriad species of insects, spiders, and other arthropods. There are more than 4,000 invertebrate species known in the Great Smoky Mountains National Park (Sharkey 2001), and easily a thousand insect species in the Coweeta basin alone. The forest environment, with its favorable microclimates and structural diversity, offers a large variety of niches, different host-plant species, and soil and litter habitats (figure 9.1). Of this vast assemblage of arthropod species, most are predators that keep prey populations at low abundances, and only a few insects ever reach population sizes that can cause any economic damage to the forest. When these occasional outbreaks do occur, they can be severe. The discipline of forest entomology has the goal of preventing timber loss to insects, and most of the knowledge of forest insects has been developed within the context of economic importance (Coulson and Witter 1984). The principal goal of insect ecology, in contrast, is to understand insect response to and influence on ecological processes (Schowalter 2011).

In the Coweeta basin, the notable death of trees has occurred due to the activities of the elm spanworm (*Ennomos subsignarius*) on hickories and oaks in 1954–1964 (Fedde 1964); the fall cankerworm (*Alsophila pometaria*) in 1972–1978, mostly on oaks (Swank et al. 1981); and the southern pine beetle (*Dendroctonus frontalis*) in 1986–1989 (Smith 1991; Kloeppel et al. 2003; Birt 2011). Other minor outbreaks have been documented, such as defoliation of oaks (*Quercus* spp.) by sawflies (*Periclista* spp.) in 1998–1999 (Reynolds et al. 2000). Southern pine beetle and sawfly outbreaks were associated with drought conditions.

A wood-boring beetle species was of special interest for the WS 7 clearcutting experiment. The locust borer (*Megacyllene robiniae*; figure 9.2) causes death of

* Corresponding author: Department of Environmental Studies, One University Heights, University of North Carolina at Asheville, Asheville, NC 28804 USA

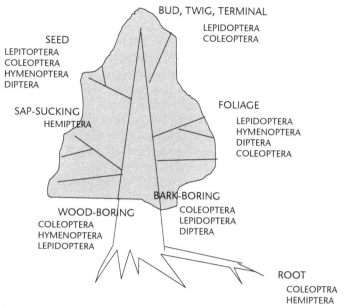

Figure 9.1 Categories of insect feeding relationships on forest trees. (Adapted from Franklin 1970)

Figure 9.2 The locust borer, *Megacyllene robiniae.* (Photo by D. Crossley)

black locust (*Robinia psuedoacacia*) trees throughout the United States (Wollerman 1970). We anticipated the proliferation of black locust on WS 7 during the course of revegetation (see Boring et al., chapter 2, this volume), as earlier we had observed the death of stands of black locust on WS 6, following an outbreak of the locust borer. Nitrogen released from the dying trees appeared in the stream draining WS 6, and growth of tulip-poplar trees on that watershed was stimulated by nitrogen released from the dead and dying black locust trees (Boring 1982; Boring and Swank 1984). We expected that the locust borer would make its appearance in the form of an outbreak on WS 7 some 20–30 years following clearcutting.

Most of our research on canopy arthropods at Coweeta has been organized around the importance of insects in forest nutrient dynamics (Crossley et al. 1988), although defoliation by arthropods may reduce water uptake and thus improve tree survival during drought (Kolb et al. 1999). Arthropods and other animals may regulate nutrient cycling by influencing the rates of nutrient uptake by vegetation, return to the forest floor, and release during decomposition (figure 9.3). In research performed as a part of the International Biological Program and the Long-Term Ecological Research program of the National Science Foundation, we measured the biomass of canopy consumers and their nutrient content (Schowalter et al. 1981; Schowalter and Crossley 1983; Crossley et al. 1988). To evaluate the relative importance of the thousands of arthropod species, we arrayed them into functional groups according to their feeding type, trophic position in the food web, and life-history characteristics. For example, spiders and predaceous beetles were grouped together. Aphids and leafhoppers (sap-sucking insects) were analyzed separately from caterpillars (chewing insects). Functional groupings allowed us to compare canopy arthropod assemblages between tree species and across watersheds (Schowalter et al. 1981;

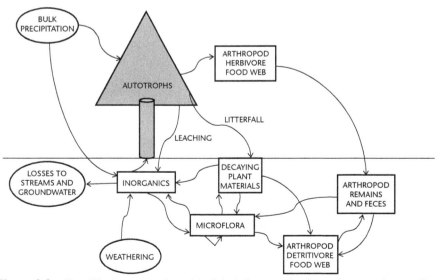

Figure 9.3   Simplified conceptual model of the influence of arthropods on nutrient cycling processes in forests. (Adapted from Seastedt and Crossley 1984)

Seastedt et al. 1983; Schowalter and Crossley 1983; Hargrove et al. 1984; Crossley et al. 1988; Schowalter and Crossley 1988; Risley and Crossley 1988; Risley and Crossley 1993; Reynolds and Crossley 1997).

The WS 7 clearcutting experiment posed some additional questions for studies of canopy arthropods. Given the expected shift in plant types (see Boring et al., chapter 2, this volume), how would the community of arthropods respond? Would the increase in herbaceous vegetation influence the community of insects on the tree vegetation? Would the regrowth foliage tissue be more tender, perhaps engendering outbreaks of defoliators? And, could we isolate those factors most responsible for the development of the insect community?

Methods for Sampling Canopy Arthropods

Modern research in forest canopies uses a variety of methods (Lowman 2004; Lowman et al. 2012), ranging from climbing techniques borrowed from mountain climbers (Schowalter 1995) to floating platforms and dirigibles (Hallé 1998). Large construction cranes have been erected in certain forests, allowing sampling of arthropods in the top of the canopy (Parker et al. 1992; Schowalter and Ganio 1998; Basset et al. 2003). Other forests have been fitted with extensive canopy walkway systems (Lowman and Bouricious 1995; Reynolds and Crossley 1995). These new approaches are allowing canopy access in unprecedented ways.

Canopy access on WS 7 was more limited. We simply used long-handled insect nets for access into lower canopies. For a few years, the low-regrowth vegetation on WS 7 could be sampled with ordinary insect nets. We fitted the insect net with a plastic bag that could be closed with a drawstring. The bag was slipped over a mass of leafy vegetation and quickly closed, capturing arthropods and the vegetation supporting them. A long pole pruner was then used to clip the branch. We expressed the density of arthropods as weight inhabiting the mass of vegetation captured in the net (table 9.1).

Table 9.1   Density of canopy arthropod functional groups on clearcut WS 7 and uncut control WS 2, means of five tree species (red maple, chestnut oak, dogwood, hickories, and tulip poplar).

	1975	1977		1978	
	WS 7 Precut	WS 2 Control	WS 7 1st year postcut	WS 2 Control	WS 7 2nd year postcut
Chewing	270.4	200.2	368.0	171.2	137.0
herbivores	±56.1	±68.6	±193.3	±57.0	± 62.0
Sucking	91.6	57.6	148.2	76.6	104.0
herbivores	±26.7	±9.9	±34.0	±8.4	±23.0
Omnivores	30.0	4.4	87.6	10.2	18.4
	±11.8	±1.3	±54.9	±2.0	±4.8
Predators	138.2	91.4	54.8	105.0	76.4
	±51.3	±12.7	±11.9	±24.8	±27.7
Totals	530.2	353.6	658.4	363.0	335.8

Note: Data are mg arthropods per kg foliage (± standard error).

The consumption of vegetation by chewing insects was estimated by measuring the holes in leaves—the leaf area missing. A digitizer and analyzer were constructed using an office copier and a small personal computer (Hargrove and Crossley 1988).

Results

Immediate Response of Canopy Arthropods to Clearcutting

The summer after WS 7 was clearcut, tree foliage returned on sprouts from stumps (see Boring et al., chapter 2, this volume). This leafy mass contained large numbers of arthropods, and their mass differed from that found before cutting or in the adjacent uncut control WS 2 (table 9.1). In comparison with WS 2, the clearcut had a greater mass of chewing herbivores, sucking herbivores, and omnivores. However, the mass of predators on the clearcut was lower. The chewing herbivores (caterpillars, beetles, crickets, leaf miners) exhibited nearly twice the mass found on WS 2. We had anticipated an increase in sucking herbivores (aphids, leafhoppers, and treehoppers) on succulent regrowth foliage. Aphids in particular increased in density on all tree species sampled, except for dogwood, the first year after cutting. The omnivores reported in table 9.1 were largely species of ants. Increases in ant density accompanied the increases in aphids (see also Crossley et al. 1988). Predators included a variety of spiders, predaceous beetles, lacewings, and wasps. By the second summer (1978) masses of chewing herbivores had declined to levels similar to those on WS 2, and masses of sucking herbivores and omnivores were only slightly elevated (table 9.1).

The data shown in table 9.1 are means for five tree species. The rather large standard errors attached to those means are due to large differences in attendant insects that were found among the tree species. For example, tulip-poplar foliage on WS 7 supported large masses of chewing insects (caterpillars and leaf-feeding beetles) and aphids, much in excess of those on WS 2. In contrast, chewing insects on dogwoods and hickories showed a smaller response (Schowalter et al. 1981).

The increases in arthropod density following cutting were not surprising. We anticipated that the regrowth foliage might be more palatable, and that opportunistic insects such as the aphids might increase in comparison with those on the control WS 2. The picture is less clear in view of the results of sampling WS 7 two years previous to the clearcut. Chewing herbivores, sucking herbivores, omnivores, and predators were all more abundant in precut WS 7 samples than on WS 2 in either of the postcut years (table 9.1). Unfortunately, no comparable samples are available from WS 2 in the precut year. Given the precut similarity of WS 2 and WS 7 in essential physical and vegetation characteristics, we would expect their precut arthropod communities to have been similar also. There is often a large year-to-year variation in insect abundance, and such annual variation may be a factor in the differences between 1975, 1977, and 1978. Furthermore, different personnel performed the sampling in 1975, and the discrepancies may reflect slight differences

Table 9.2   Density of canopy arthropods on clearcut WS 7 and uncut control WS 2, means of five tree species (red maple, chestnut oak, dogwood, hickories, and tulip poplar).

	1984		1985		1987	
	WS 2 Control	WS 7 Clearcut	WS 2 Control	WS 7 Clearcut	WS 2 Control	WS 7 Clearcut
Mean density	559.7	397.9	349.8	337.2	268.8	653.4
(Standard error)	(117.9)	(113.1)	(124.9)	(56.8)	(72.0)	(249.5)

Note: Data are mg arthropods per kg foliage (± standard error).
Source: Blanton (1989).

in sampling technique. In any case, the first postcut year (1977) had higher insect densities than the precut 1975 samples or concurrent samples from WS 2.

Further sampling of canopy arthropods in 1984–1987 was undertaken by Blanton (1989). She reported only total arthropod weight densities, not broken down into functional groups (table 9.2). During this period, a decade after the clearcut, there was no significant difference in arthropod mass density between WS 7 and WS 2. The years 1984 and 1985 were characterized by reduced rainfall in the Coweeta basin. Blanton noted that, during 1985, chewing herbivores decreased during this drought period and that sucking herbivores (aphids, membracids, psyllids) increased. Precipitation returned to more normal levels in 1987 and the proportion of chewing herbivores increased. Thus, the response of arthropod functional groups to a two-year drought resembled their response to clearcutting.

Black Locust: A Special Case

Black locust is an important tree species on successional watersheds because of its ability to fix atmospheric nitrogen in the rhizosphere. On WS 7, black locust was home to a large array of arthropod species. Large numbers of aphids (*Aphis craceivora*) and membracids (mostly *Vanduzea arquata*) were tended by colonies of ants (*Formica integra*) (Schowalter et al. 1981; Hargrove 1986; see figure 9.4). In the springtime the ants gathered in clusters at the tips of the tree branches, possibly feeding upon extrafloral nectaries. Chewing herbivores were well represented also, including geometrid caterpillars, the silver-spotted skipper (*Epargyrus clarus*) and numbers of flea beetles (*Derocrepis carinata*) (Hargrove 1986).

The density of arthropod functional groups measured on black locust was higher than the average density for arthropods on other tree species on WS 2 and WS 7 (table 9.3). Following clearcutting, densities of sucking herbivores increased dramatically over those measured during the precut year (1975) or on control WS 2. The omnivore category, largely ants, also increased markedly in the two postcut years. Blanton (1989) measured total arthropod density for black locust on WS 7 and found that density remained high in 1984 (1390 mg per kg foliage) and 1985 (1355 mg per kg), values similar to those reported in table 9.2 for the immediate postcut years.

Figure 9.4   Ants tending aphids. (Photo by T. Schowalter)

Table 9.3   Density of canopy arthropod functional groups on black locust foliage.

	1975	1977		1978	
	WS 7 Precut	WS 2 Control	WS 7 1st year postcut	WS 2 Control	WS 7 2nd year postcut
Chewing	956	414	851	359	251
herbivores	±200	±270	±331	±126	±70
Sucking	180	71	567	172	1486
herbivores	±30	±34	±264	±58	±561
Omnivores	7	25	149	64	227
	±4	±13	±62	±22	±75
Predators	164	57	77	96	48
	±72	±18	±36	±55	±17
Totals	1307	567	1644	691	2012

Note: Data are mg arthropods per kg foliage (± standard error).
Sources: 1975 data, Petursson (unpublished); 1977 and 1978 data, Schowalter et al. (1981).

Black locust foliage showed the effects of consumption by chewing herbivores. Hargrove (1983) estimated the amount of leaf area removed by reconstructing damaged leaflets and measuring the area with a digitizer (Hargrove and Crossley 1988), and showed that damage to black locust foliage accumulated during the summer (table 9.4). Leaf area missing due to insect feeding amounted to an average of 15% by September. Correspondingly, the numbers of undamaged leaflets declined during the summer. On the average, fewer than 10% of the leaflets showed no insect damage in September.

Table 9.4   Estimates* of leaf area consumed on black locust trees, WS 7, summer, 1980.

	Month			
	June	July	August	September
Mean leaf area consumed	4.8%	8.1%	13.2%	15.3%
(Range of values)	(2.8–6.2)	(4.8–11.4)	(8.1–16.1)	(7.3–24.7)
Percent of leaflets undamaged	50.8%	42.2%	15.3%	7.3%
(Standard error)	(6.88)	(6.61)	(3.26)	(2.58)

*Values represent means for six trees. Each tree sample consisted of 75 leaflets
Source: Adapted from Hargrove (1983)

Accumulation of damage in this manner is unusual for forest trees. Most measurements of holes in canopy leaves show that 3%–8% of leaf area is missing, and that area does not increase during the season (Risley 1987). Several factors may account for the lack of damage accumulation. In general, canopies have a burst of insect activity in May and June, and then little in midsummer, followed by another increase in herbivory in late August and September (Blanton 1989). For example, the most chewing consumption on chestnut oak occurred in June. Some tree species have an additional flush of leaves during the summer. Further, partially damaged leaves may abscise (Risley 1987). These considerations suggest that our estimates of herbivory on black locust may be low. But it appears from table 9.4 that little regrowth of foliage occurs on black locust and abscission of damaged leaflets would seem to be a minor factor.

To summarize, regrowth tree foliage one year after the clearcut had a greater mass of chewing herbivores, sucking herbivores, and omnivores, such as caterpillars, aphids, and ants, respectively, compared to the adjacent control watershed. Predators, such as spiders and predaceous beetles, decreased. Differences for black locust were especially marked, with greater densities seen especially for sucking herbivores and omnivores, particularly ants. Measurements of leaf area removed on black locust averaged 15% by September, a fairly high value. However, no obvious outbreaks of defoliators were observed.

Literature Cited

Basset, Y., V. Horlyck, and S. J. Wright. 2003. Studying forest canopies from above: the International Canopy Crane Network. Smithsonian Tropical Forest Research Institute and UNEP, Panama.

Birt, A. 2011. Regional population dynamics. Pages 109–128 in Southern pine beetle II. R. N. Coulson and K. D. Kepzig, editors. USDA Forest Service, Southern Research Station, Asheville, North Carolina.

Blanton, C. M. 1989. Canopy arthropod communities in the southern Appalachians: Impacts of forest management and drought. PhD dissertation. University of Georgia, Athens.

Boring, L. R. 1982. The role of black locust (*Robinia pseudoacacia* L.) in forest regeneration and nitrogen fixation in the southern Appalachians. PhD dissertation. University of Georgia, Athens.

Boring, L. R, and W. T. Swank. 1984. The role of black locust (*Robinia pseudo-acacia*) in forest succession. *Journal of Ecology* 72: 749–766.

Crossley, D. A., Jr., C. S. Gist, W. W. Hargrove, L. S. Risley, T. D. Schowalter, and T. R. Seastedt. 1988. Foliage consumption and nutrient dynamics in canopy insects. Pages 193–206 in *Forest Hydrology and Ecology at Coweeta*. W. T. Swank and D. A. Crossley, Jr., editors. Springer-Verlag, New York, New York.

Coulson, R. N., and J. A. Witter. 1984. *Forest Entomology: Ecology and Management*. Wiley Interscience, New York, New York.

Fedde, G. F. 1964. Elm spanworm, a pest of hardwood forests in the southern Appalachians. *Journal of Forestry* 62: 102–106.

Franklin, R. T. 1970. Insect influences on the forest canopy. Pages 86–99 in *Analysis of Temperate Forest Ecosystems*. D. E. Reichle, editor. Springer-Verlag, New York, New York, USA.

Hallé, F., editor. 1998. Biologie D'une Canopée de Forêt Equatoriale III. Pro-Natura Interntional & Opération Canopée, France.

Hargrove, W. W. 1983. Forest canopy consumption by arthropod herbivores: an average availability model. MS thesis. University of Georgia, Athens.

Hargrove, W. W. 1986. An annotated species list of insect herbivores commonly associated with black locust, *Robinia pseudoacacia*, in the southern Appalachians. *Entomological News* 97: 36–40.

Hargrove, W. W., and D. A. Crossley, Jr. 1988. Video digitizer for the rapid measurement of leaf area lost to herbivorous insects. *Annals of the Entomological Society of America* 81: 593–598.

Hargrove, W. W., D. A. Crossley, Jr., and T. R. Seastedt. 1984. Shifts in insect herbivory in the canopy of black locust, *Robinia pseudoacacia*, after fertilization. *Oikos* 43: 322–328.

Kloeppel. B. C., B. D. Clinton, J. M. Vose, and A. R. Cooper. 2003. Drought impacts on tree growth and mortality of southern Appalachian forests. Pages 43–55 in *Climate Variability and Ecosystem Response at Long-Term Ecological Research Sites*. D. Greenland, D. G. Goodin, and R. C. Smith, editors. Oxford University Press, New York, New York.

Kolb, T. E., K. A. Dodds, and K. M. Clancy. 1999. Effect of western spruce budworm defoliation on the physiology and growth of potted Douglas-fir seedlings. *Forest Science* 45: 280–291.

Lowman, M. D. 2004. Tarzan or Jane? A short history of canopy biology. Pages 453–464 in *Forest Canopies*. M. D. Lowman and H. B. Rinker, editors. Elsevier Press, Amsterdam, the Netherlands.

Lowman, M. D., and B. Bouricious. 1995. The construction of platforms and bridges for forest canopy access. *Selbyana* 16: 179–184

Lowman, M. D., T. D. Schowalter, and J. F. Franklin. 2012. *Methods in Forest Canopy Research*. University of California Press, Berkeley, California.

Parker, G., A. P. Smith, and K. P. Hogan. 1992. Access to the upper canopy with a large tower crane. *BioScience* 42: 664–671.

Reynolds, B. C. and D. A. Crossley, Jr. 1995. Use of a canopy walkway for collecting arthropods and assessing leaf area removed. *Selbyana* 16: 21–23.

Reynolds, B. C., and D. A. Crossley, Jr. 1997. Spatial variation in herbivory by forest canopy arthropods along an elevation gradient. *Environmental Entomology* 26: 1232–1239.

Reynolds, B. C., M. D. Hunter, and D. A. Crossley, Jr. 2000. Effects of canopy herbivory on nutrient cycling in a northern hardwood forest in western North Carolina. *Selbyana* 21: 74–78.

Risley, L. S. 1987. Acceleration of seasonal leaf fall by herbivores in the southern Appalachians. PhD dissertation. University of Georgia, Athens.

Risley, L. S., and D. A. Crossley, Jr. 1988. Herbivore-caused greenfall in the southern Appalachians. *Ecology* 69: 1118–1127.

Risley, L. S., and D. A. Crossley, Jr. 1993. Contributions of herbivore-caused greenfall to litterfall nitrogen flux in several southern Appalachian watersheds. *American Midland Naturalist* 129: 67–74.

Schowalter, T. D. 1995. Canopy arthropod community responses to forest age and alternative harvest practices in western Oregon. *Forest Ecology and Management* 78: 115–125.

Schowalter, T. D. 2011. *Insect Ecology: An Ecosystem Approach*. 3rd ed. Elsevier/Academic, San Diego, California.

Schowalter, T. D., and D. A. Crossley, Jr. 1983. Forest canopy arthropods as sodium, potassium, magnesium and calcium pools in forests. *Forest Ecology and Management* 7: 143–148.

Schowalter, T. D., and D. A. Crossley, Jr. 1988. Canopy arthropods and their response to forest disturbance. Pages 207–219 in *Forest Hydrology and Ecology at Coweeta*. W. T. Swank and D. A. Crossley, Jr., editors. Springer-Verlag, New York, New York.

Schowalter, T. D., and L. M. Ganio. 1998. Vertical and seasonal variation in canopy arthropod abundances in an old-growth conifer forest in southwestern Washington. *Bulletin of Entomological Research* 88: 633–640.

Schowalter, T. D., J. W. Webb, and D. A. Crossley, Jr. 1981. Community structure and nutrient content of canopy arthropods in clearcut and uncut forest ecosystems. *Ecology* 62: 1010–1019.

Seastedt, T. R., and D. A. Crossley, Jr. 1984. The influence of arthropods on ecosystems. *BioScience* 34: 157–161.

Seastedt, T. R., D. A. Crossley, Jr., and W. W. Hargrove. 1983. The effects of low-level consumption by canopy arthropods on the growth and nutrient dynamics of black locust and red maple trees in the southern Appalachians. *Ecology* 64: 1040–1048.

Sharkey, M. J. 2001. The all taxa biological inventory of the Great Smoky Mountains National Park. *Florida Entomologist* 84: 556–564.

Smith, R. N. 1991. Species composition, stand structure, and woody detrital dynamics associated with pine mortality in the southern Appalachians. MS thesis. University of Georgia, Athens..

Swank, W. T., J. B. Waide, D. A. Crossley, Jr., and R. L. Todd. 1981. Insect defoliation enhances nitrate export from forest ecosystems. *Oecologia*. 51: 297–299.

Wollerman, E. H. 1970. The locust borer. U.S. Department of Agriculture Forest Service Forest Pest Leaflet 71.

10

Recovery of Particulate Organic Matter Dynamics in a Stream Draining a Logged Watershed

A Pressing Situation

Jackson R. Webster*
E. F. Benfield
Stephen W. Golladay
Matthew E. McTammany

Introduction

Compared to other ecosystems, streams have been described as having low resistance to disturbance but being highly resilient following disturbance (Webster et al. 1975). Webster et al. (1983) demonstrated the low resistance of Big Hurricane Branch at Coweeta to watershed logging, documenting the many changes to the stream in the first few years after clearcutting. These changes were associated with physical disturbance (high sediment export), increased nutrient export, changes related to the reduction of the forest canopy (decreased allochthonous inputs, increased light, increased autochthonous production), and alterations in the invertebrate community. Although there had not been sufficient time to see resilience, Webster et al. (1983) noted that within 4 years following disturbance some aspects of the stream were showing significant return to pre-logging conditions. They also noted that some of the changes in the stream, particularly the shift to an autochthonous energy base and the macroinvertebrate community shift from shredder- to grazer-dominated, could be interpreted as mechanisms of resilience. They concluded that the potential resilience of the stream could not be realized because of the long-term modification of the quantity and quality of terrestrial organic matter inputs (Webster and Patten 1979; Gurtz et al. 1980).

*Corresponding author: Department of Biological Sciences, Virginia Polytechnic Institute and State University, Blacksburg, Virginia 24061

After several years of further study, Webster et al. (1992) predicted a postdisturbance period of accelerated sediment loss due to a decline in large wood in the stream. Because the logs existing in the stream prior to logging and those introduced during logging decay slowly, the regrowing forest will not provide new large wood, and it might be from 50 to 200 years before logs of sufficient size to form stable dams fall into the stream. However, they also pointed out that decay-resistant large wood from earlier disturbances could modify this trend. Wallace et al. (2001) found that chestnut logs resulting from the chestnut blight in the 1930s constituted 24% of the large wood in a small Coweeta stream. These large, old, decay-resistant logs are abundant in Big Hurricane Branch and may provide a bridge between logging and new inputs of large wood. The hemlock woolly adelgid (*Adelges tsugae*) is now present at Coweeta and throughout most of the Appalachian forests (Webster et al. 2012) and may cause a similar input of very large and even more decay-resistant wood to streams in the near future.

The concepts of press and pulse disturbances were introduced by Bender et al. (1984) to describe various types of community ecology experiments, but these terms can also be applied in a much broader context. Ecosystem stability theory, as adapted from mathematical theory, was based on short pulse disturbance (Rosenzweig and MacArthur 1963; Waide and Webster 1976). Pulse disturbance models may be broadly applicable to many situations; for example, following logging, the terrestrial ecosystem (viewed as the whole system within the physical space affected by the logging) is free to return to its original condition. The external factors that modified this system have returned to what they were before the disturbance; that is, the forest is no longer being logged. However, many ecological disturbances are not pulses, but continue for a long time relative to the potential rate of recovery of the system. These can be described as press disturbances. This is the case for the effect of logging on streams. Streams are tightly linked to their terrestrial surroundings (e.g., Hynes 1975; Vannote et al. 1980), and as long as this linkage is modified, the press disturbance continues. Specifically, as long as solar input, allochthonous inputs, hydrologic regime, and other terrestrial-stream linkages remain modified, a stream draining a logged forest continues to be disturbed. Many aspects of the potential resilience of the stream ecosystem cannot be realized, as the stream continually tracks a long-term press disturbance (Webster and Patten 1979).

Watershed (WS) 7 at Coweeta was logged in 1977. The stream draining this watershed, Big Hurricane Branch, was affected in many ways (Gurtz et al. 1980; Webster et al. 1983). While the stream has recovered in some characteristics, the continuing press disturbance limits many aspects of recovery. In this chapter, we report the long-term pattern of recovery of the organic matter dynamics of this stream.

Site Description

The study was conducted at the Coweeta Hydrologic Laboratory in western North Carolina. Coweeta is a US Forest Service Experimental Field Station and a Long-Term Ecological Research site supported by the National Science Foundation and the US Forest Service. The Coweeta basin is organized into multiple experimental watersheds, which have been subjected to various whole-watershed manipulations over the last 70 years. There are also a number of long-term reference watersheds that have not been disturbed since being logged prior to 1922, with the exception of the American chestnut (*Castanea dentata*) blight in the 1930s. The forest is dominated by oaks (*Quercus* spp.), red maple (*Acer rubrum*), hickories (*Carya* spp.), and yellow poplar (*Liriodendron tulipifera*). Streamside vegetation consists of many hemlocks (*Tsuga canadensis*), birch (*Betula* spp.), and an often dense subcanopy of rhododendron (*Rhododendron maximum*). In-stream primary production is extremely low (Mulholland et al. 1997), and the benthic community is dominated by insects that are leaf shredders or particle collectors (see Wallace and Ely, chapter 11, this volume).

In this chapter, we report the on the recovery of Big Hurricane Branch from watershed logging, comparing our results to a reference stream, Hugh White Creek. Characteristics of the two streams and their watersheds are summarized in table 10.1. Big Hurricane Branch drains WS 7, which was clearcut in 1977 (figure 10.1). Efforts were made to minimize disturbance to the soil surface by lifting logs from steep slopes to roads by an aboveground cable system. However, tractor skidding was employed on gentler slopes. Three major roads crossing the main stream and tributaries of Big Hurricane Branch were constructed in the basin. Following logging, large slash was removed from much of the stream channel. Prior to logging in 1977, the lower part of the basin had been experimentally used for cattle grazing for a few months each year from 1941 to 1952 and was logged prior to 1920. The basin was undisturbed from 1952 until the logging operation in 1977. After logging, regrowth vegetation was dominated by herbaceous plants, black locust, and rapidly sprouting species, such as red maple (Boring et al. 1981; see figure 10.2). Current watershed vegetation is dominated by sprout regrowth hardwood trees (see Boring et al., chapter 2, this volume). The response of this stream to watershed logging has been described in several papers (e.g., Gurtz et al. 1980; Webster et al. 1983; Gurtz and Wallace 1984; Webster et al. 1992)

Hugh White Creek drains WS 14, one of the long-term reference watersheds at Coweeta. WS 14 was logged in the early 1920s but has since been undisturbed. This stream has been extensively studied and further description of the stream and references to earlier studies were published by Webster et al. (1997).

Table 10.1   Physical characteristics of the two study streams.

Variable	Big Hurricane Branch	Hugh White Creek
Watershed	7	14
Treatment	Clearcut	Long-term reference
Watershed area (ha)	58.7	61.1
Mainstream length (m)	1225	1125
Gradient (m/m)	0.19	0.15
Mean annual discharge (L/s)	18.5	19.4
Max watershed elev. (m)	1060	996
Min watershed elev. (m)	724	708
Basin orientation	South facing	North facing

Figure 10.1 View of lower section of Big Hurricane Branch (WS 7), in the first year after cutting in 1977. (Photo by J. Webster)

Figure 10.2 View of lower section of Big Hurricane Branch (WS 7) in 1982, five years after cutting. (Photo by J. Webster)

Methods

Litterfall

Annual litter inputs to Big Hurricane Branch were measured just prior to logging and have been measured six times since logging. Similar measurements of litterfall to Hugh White Creek were made in 1983–84, 1993–94, and 2003–2004. All measurements were made using overstream or streamside baskets or traps. Collection sites were located along the length of the main channel of each stream. In general, litter falling in the baskets was collected twice monthly from September through November and then monthly through the rest of the year. In the laboratory, litter was separated into leaves, wood, and miscellaneous material, leaves were separated to species, and all material was air-dried to a constant weight and weighed. Subsamples were ashed at 500°C to determine ash free dry mass (AFDM).

Leaf Breakdown

Leaf breakdown was measured in Big Hurricane Branch before logging and seven times subsequently. Concurrent measurements of leaf breakdown in Hugh White Creek were begun in 1983. All measurements were made using the mesh bag technique (Benfield 2006). Senescent leaves were collected at Coweeta just prior to abscission and weighed amounts were placed in mesh bags (3–5 mm openings). Bags were placed in the streams in late autumn and either anchored to the stream bottom with large nails or tied to roots or small trees. Replicate bags of each species were taken back to the lab to account for weight loss due to handling. Three to five bags of each species from each site were retrieved periodically, returned to the laboratory, rinsed to remove sediment, invertebrates, and debris, and then air-dried to a constant weight and weighed. Subsamples were ashed at 500°C to determine AFDM. Breakdown rates were determined by regressing natural log of mass remaining versus time (e.g., Webster and Benfield 1986). For two studies, we also determined breakdown rates by regressing log mass remaining versus cumulative degree-days (e.g., Minshall et al. 1983).

Benthic Standing Stock of Particulate Organic Matter

Leaves

Leaf standing stock on the streambed was sampled prior to logging and six times subsequently. In the first two studies, leaves were sampled using a Surber sampler; in later studies samples were taken with a core sampler or Surber sampler in most areas and with a core in depositional areas. In general, coarse particulate organic material was removed by hand and placed in a 1-mm mesh net and rinsed to remove smaller material. In the laboratory, the material was separated into leaves, wood, and other (nuts, flowers, etc.), air-dried to constant weight, and weighed. Subsamples were ashed at 500°C to determine AFDM.

Fine Benthic Organic Matter (FBOM)

Fine benthic organic matter (FBOM) was measured along with the core measurements of larger particles in 1985–86 (Golladay et al. 1989) and again in July 1994 (Webster and Benfield, unpublished). Water from the core was pumped through a 1-mm mesh net until the water was clear. We then measured the volume of water and collected an approximately 1-L sample. The volume of sample was measured in the laboratory, and a measured subsample was filtered though a preweighed glass fiber filter (mesh opening approximately 0.5 μm). We determined the AFDM of material on the filter (Golladay et al. 1989) and used the data to estimate FBOM standing crop in the stream.

Wood

Small wood was sampled in the Surber and core samples, and larger wood was estimated either by measuring all the wood in 1-m transects or by the line-intersect method. Using the transect method, all medium-sized (1–5 cm diameter) wood was collected and weighed. Subsamples of the collected wood were wet-weighed in the field, and the rest of the material was returned to the stream. In the laboratory, the subsamples were dried, weighed, ashed, and reweighed. Data from the subsamples were used to convert field-measured wet weights to AFDM. All individual pieces of larger wood (> 5 cm diameter) were either wet-weighed, subsampled, and treated as we did the smaller pieces; or, for very large logs, we measured the length and diameter and collected subsamples with a hand saw. The subsamples of these logs were dried and weighed; and the volume was measured by water displacement. Subsamples were then taken for determination of AFDM, and the measurements were used to determine the AFDM of the logs.

In 1995 (Big Hurricane Branch) and 1999 and 2008 (Big Hurricane Branch and Hugh White Creek), wood was also measured by the line-intersect method (Warren and Olsen 1964; Wallace and Benke 1984). In 1995 and 1999 a line was placed across the stream at each of 60 cross-sections and the diameter of all wood intersecting this line was measured. In 2008 a line was placed along the thalweg of each stream. The measurements were converted to volume and then to AFDM using density measurements from the transect study or a general value of 0.4 g/cm^3.

Simulations of Leaf Dynamics

Using the leaf input and breakdown rates described earlier, we developed a computer model to predict leaf standing stocks, which we then compared against measured values. The initial model had the following form:

$$\frac{dX_i}{dt} = I_i(t) - k_i X_i$$

where X_i is leaf standing stock, t is time, $I_i(t)$ is leaf-fall, k_i is the exponential breakdown rate, and i is the leaf-breakdown-rate category. We used breakdown rates for four categories of leaves, fast, medium, slow, and very slow, based on the leaf-breakdown measurements made in 1993–94 for Big Hurricane Branch and the averages of all rates measured at Coweeta in reference streams for Hugh White Creek (Webster et al. 1999; Webster et al. 2001). Daily leaf-fall rates were calculated by linear interpolation from data collected in 1993–94. We did not include loss via transport in the model. Previous studies at Coweeta have shown that streams of this size are highly retentive, leaves breakdown very close to where they enter the stream, and leaf export is a small fraction of the total leaf input (e.g., Webster et al. 1999). The model was solved numerically using Runge-Kutta integration with a FORTRAN computer program.

We subsequently made two modifications to this initial model. First, we included blow-in, the lateral movement of leaves into the stream. Blow-in was based on measurements made in 1983–84 (Webster et al. 1990) modified to the leaf composition measured in 1993–94. Second, to include temperature, we modified the model to use degree-day breakdown rates and mean daily water temperatures.

Results

Water Temperature

During the first few years following logging, summer water temperature in Big Hurricane Branch was elevated due to the absence of a forest canopy (Swift 1983; Webster et al. 1983). However, the temperature elevation only lasted a few years. Once the forest canopy closed, water temperature in Big Hurricane Branch showed less annual variation than in Hugh White Creek (Stout et al. 1993). For example, during 1993–94 and again in 2004–2005, water temperature in Big Hurricane Branch was warmer in winter and cooler in summer than in Hugh White Creek (figure 10.3). This difference is apparently due to geomorphological differences of the two watersheds and not to logging. Annual degree-days were slightly higher in Big Hurricane Branch during both of these years.

Leaf-fall

Prior to logging, leaf-fall to Big Hurricane Branch was 259.2 g m^{-2} y^{-1}, slightly less than leaf-fall in Hugh White Creek (table 10.2; figure 10.4) and other reference streams at Coweeta (Webster et al. 1990). Leaf-fall was reduced to almost zero after logging but recovered to reference levels within five years (figure 10.4). Despite quantitative recovery, leaf-fall remained qualitatively different and was generally a mix of more labile leaf species than leaf-fall to the reference stream through 1993–94 (table 10.2). The near absence of oak leaves even in 1993–94 is particularly evident. The temporal pattern of leaf-fall was very similar in Big Hurricane Branch and Hugh White Creek in 1993–94, with a slightly earlier peak in Big Hurricane Branch (figure 10.5). More recently (2003–2004), the composition of leaves falling in Big Hurricane Branch has been more similar to Hugh White Creek except for the low abundance of hemlock needles in Big Hurricane Branch.

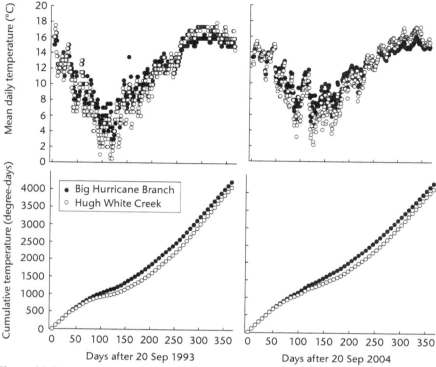

Figure 10.3   Comparison of stream water temperatures in Big Hurricane Branch and Hugh White Creek, 1993–94 and 2004–2005. Data from USDA Forest Service, Coweeta; and Webster and Benfield (unpublished).

Leaf Breakdown Rates

Several patterns are evident in the breakdown rates (figure 10.6). First, leaf breakdown rates in Big Hurricane Branch before logging were very similar to rates in Hugh White Creek. Immediately following logging, breakdown rates were slower, probably due to sediment burial of the leaf packs (Webster and Waide 1982). However, subsequent measurements indicate breakdown rates significantly faster than pretreatment rates and faster than rates measured in Hugh White Creek. Across both species and streams, the annual patterns are nearly identical—rates were fairly fast in 1983–84, slower in 1986–87, faster again in 1995–96, and slower again in 1999–2000. These annual trends are probably related to differences in winter water temperatures or annual discharge (Benfield et al. 2001). However, in the most recent measurements (2005–2006), breakdown rates of rhododendron, white oak, and red maple in Hugh White Creek were slightly faster than measurements made 6 years previously while breakdown rates of these leaf species were slightly slower in Big Hurricane Branch. Leaf breakdown rates calculated on a degree-day basis (table 10.3) show the same trend, that is, substantially faster breakdown of leaves in Big Hurricane Branch even when corrected for temperature.

Table 10.2 Total leaf-fall and species composition of leaf-fall to Big Hurricane Branch (BHB) and Hugh White Creek (HWC).

Species	Percent of annual leaf-fall						
	BHB 1974–75	BHB 1978–79	BHB 1983–84	BHB 1993–94	HWC 1993–94	BHB 2003–04	HWC 2003–04
Birch (*Betula* spp.)	5.1	15.7	11.8	11.6	24.0	11.3	17.5
Rhododendron (*Rhododendron maximum*)	11.6	26.5	11.6	14.3	18.2	23.5	21.3
Yellow poplar (*Liriodendron tulipifera*)	9.1	10.8	2.2	13.8	14.5	23.0	16.1
White oaks (*Quercus alba, Q. prinus*)	19.0	—	3.0	—	9.0	2.1	0.4
Hemlock (*Tsuga canadensis*)	—	—	—	—	6.8	1.7	13.5
Hickory (*Carya* spp.)	11.4	—	—	—	6.4	—	—
Red maple (*Acer rubrum*)	4.8	—	11.0	7.8	5.9	5.1	3.0
Red oak (*Q. rubra*)	13.4	—	2.5	—	4.4	3.9	3.4
Basswood (*Tilia americana*)	4.2	—	—	—	—	—	—
Dogwood (*Cornus florida*)	1.1	3.9	6.5		—	—	—
Black locust (*Robinia pseudoacacia*)	—	—	3.1	2.8	—	—	—
Beech (*Fagus grandifolia*)	7.9	—	1.9	5.4	—	—	—
Ash (*Fraxinus* spp.)	—	—	—	3.8	—	—	—
Magnolia (*Magnolia* spp.)	—	—	5.0	7.6	—	—	—
Willow (*Salix nigra*)	—	—	4.3	—	—	—	—
Others*	12.5	43.1	37.1	33.7	9.3	29.2	24.1
Total leaf-fall (gAFDM/m²/y)	259.2	4.2	354.2	342.2	327.0	292.4	305.7

*'Others' category includes all species making up less than 2% of annual leaf-fall.
Sources: Data from Webster and Waide (1982); Webster et al. (1983); Webster et al. (1990); Webster et al. (2001); and Webster and Benfield (unpublished).

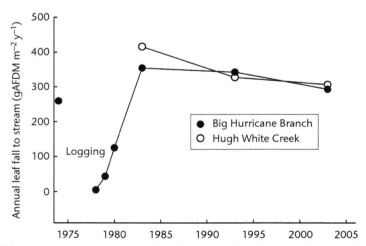

Figure 10.4 Long-term recovery of leaf inputs to Big Hurricane Branch. Data from Webster and Waide (1982); Webster et al. (1983); Webster et al. (1990); and Webster and Benfield, unpublished.

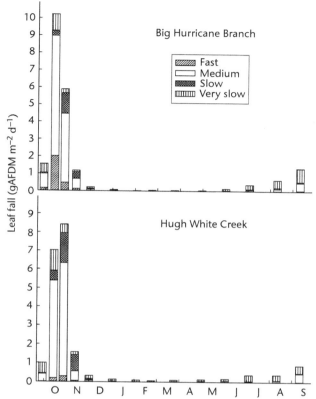

Figure 10.5. Comparison of leaf-litter inputs to Big Hurricane Branch and Hugh White Creek, 1993–94. Data from Webster and Benfield, unpublished.

Leaf Standing Stock

Interpretation of the differences of leaf standing crop between streams is difficult because of differences in sampling techniques among studies (table 10.4). However, the more intensive studies (1985–86 and 1993–94) demonstrated lower standing crop in Big Hurricane Branch than in Hugh White Creek. The difference between streams was not due to quantitative differences in input as pointed out earlier. For example, in 1993–94 and again in 2004–2005, the standing crops of leaves in the two streams were very similar just after leaf-fall, but in Big Hurricane Branch leaf standing crop declined much more rapidly through winter, spring, and early summer before increasing in late summer (figure 10.7).

Fine Benthic Organic Matter

The standing crop of FBOM in Big Hurricane Branch in 1985–86 was lower than in Hugh White Creek, both on an annual average and in July (table 10.5). Nine years later, FBOM in Big Hurricane Branch was still lower, though the difference was not statistically significant.

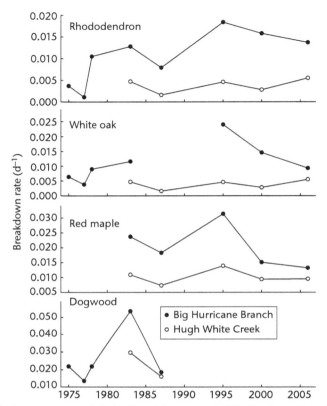

Figure 10.6   Long-term trends in leaf breakdown rates in Big Hurricane Branch and Hugh White Creek. Data from Webster and Waide (1982); Golladay and Webster (1988); Benfield et al. (1991, 2001); and Benfield and Webster, unpublished.

Table 10.3   Degree-day adjusted leaf breakdown rates* (degree-day^{-1}) in Big Hurricane Branch (BHB) and Hugh White Creek (HWC).

Variable	Big Hurricane Branch	Hugh White Creek
1994–95		
Red maple	0.00347 (0.00032)	0.00157 (0.00016)
White oak	0.00256 (0.00019)	0.00129 (0.00017)
Rhododendron	0.00194 (0.00023)	0.00054 (0.00016)
1999–2000		
Red maple	0.00132 (0.00037)	0.00102 (0.00015)
White oak	0.00171 (0.00047)	0.00082 (0.00008)
Rhododendron	0.00160 (0.00047)	0.00029 (0.00003)

*Each rate is the mean (standard error) of rates measured at three (1999–2000 and BHB 1994–95) or four (1994–95 HWC) sites in each stream.
Source: Data from Benfield and Webster (unpublished).

Table 10.4. Benthic leaf standing crop (gAFDM/m²) in Big Hurricane Branch (BHB) and Hugh White Creek (HWC).

Year of sampling	Big Hurricane Branch	Hugh White Creek	Sampling methods	References and notes
1974–76	86.2 (10.7)	—	Ten mid-stream Surber samples monthly for 2 y. Mean (standard error) of 27 sampling dates.	Webster et al. (1983); probably low because of midstream sampling.
1977–78	54.7 (11.0)	33.2 (4.5)	Weighted means of 16 Surber samples monthly for 21 mo, stratified over four habitat types. Mean (standard error) of seven seasonal values.	Gurtz (1981); Gurtz and Wallace (1984); Webster et al. (1983). HWC only sampled in second-order reach with substantial bedrock substrate.
1985–86	124 (17)	213 (18)	Seasonal sampling of 60 core (0.071 m²) samples on transects stratified along the stream. Mean (standard error) of 20 transects.	Golladay et al. (1989)
1986–87	59.9 (16.1)	102.5 (21.4)	Monthly core samples in headwater tributaries. Mean (standard error) of 11 sampling dates.	Stout et al. (1993)
1993–94	83.7 (8.2)	119.6 (18.7)	One hundred core samples on transects stratified over distance of stream, six sampling dates. Mean (standard error) of five transects.	Webster et al. (2001) and Webster and Benfield (unpublished)
1993–94	35.4	28.2	Weighted means of nine Surber or core samples collected every 2 mo for 1 y on four habitat types	Stone and Wallace (1998)
2004–05	43.7	56.2	Means of 30 core samples collected seasonally in ten transects for each stream.	Webster and Benfield (unpublished)

Wood Standing Crop

Differences in wood standing crops are also difficult to interpret because of differences in sampling techniques (table 10.6). Also, measurements of large wood can be highly biased by the presence of a single large log. Measurements of large wood in Big Hurricane Branch in 1995 were very high because of the presence of one massive log near the headwaters of this stream. However, measurements made at the same cross-sections of Big Hurricane Branch in 1995 by transect and line-intersect methods were almost identical, strengthening the case for the much simpler line-intersect method (Wallace et al. 2001). Since we began sampling large wood in 1995 there has been a substantial decrease in the amount of wood larger than 20-cm diameter (table 10.6). Though slash was removed from some of the stream, it was left in other reaches, and these data undoubtedly represent the decay of this slash, decay of residual material from before logging, and the lack of recruitment into the stream. Small wood

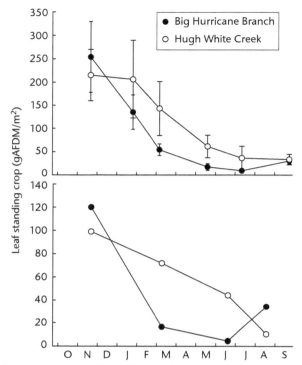

Figure 10.7 Benthic leaf material in Big Hurricane Branch and Hugh White Creek in 1993–94 (*upper panel*) and 2004–2005 (*lower panel*). The error bars for 1993–94 are 95% confidence limits using the means of the four samples from each cross section as replicates (N = 25). Data from Webster et al. (2001); and Webster and Benfield, unpublished.

Table 10.5. Fine benthic organic matter (FBOM, gAFDM/m^2) in Big Hurricane Branch and Hugh White Creek.

	Big Hurricane Branch	Hugh White Creek
Annual average, 1985–1986	112.8 (240, 7.6)	165.8 (240, 10.1)
July 1985	27.0 (27, 4.5)	76.2 (36, 9.7)
July 1994	50.2 (24, 10.7)	94.4 (24, 33.4)

Note: Data are means (number of samples, standard errors) of all samples.
Sources: Data from Golladay et al. (1989); and Webster and Benfield (unpublished).

(1–5 cm diameter) has remained very stable and similar to reference stream levels, suggesting that decay of this material is balanced by inputs of twigs in this size range.

Leaf Simulations

Our simulations of leaf standing crops in the reference stream, Hugh White Creek (figures 10.8–10), provide general support for the prediction of standing crop from measurements of input and breakdown (Webster et al. 2001). In this stream, the

Table 10.6 Wood (gAFDM/m^2) in Big Hurricane Branch (BHB) and Hugh White Creek (HWC).

Year of sampling	Diameter	Big Hurricane Branch	Hugh White Creek	Sampling methods	References and notes
1974–76	Small wood	27.0	—	Ten mid-stream Surber samples seasonally for 1 y.	Webster et al. (1983)
1977–78	Small wood	123.1	35.6	Weighted means of 16 Surber samples monthly for 21 mo, stratified over four habitat types.	Gurtz (1981); Gurtz and Wallace (1984); Webster et al. (1983)
1985–86	1–5 cm	383 (69)	312 (66)	All wood measured in 1-m transects along the stream. Mean (standard error) of 20 transects.	Golladay et al. (1989)
	>5 cm	2,833 (1,108)	5,134 (2,011)		
1993–94	<2 cm	105.5 (17.9)	106.6(19.0)	One hundred core samples on transects stratified along stream, 6 sampling dates. Mean (standard error) of means in five transects.	Webster, Benfield, Hutchens, and Tank (unpublished)
1993–94	Small wood	107.6	62.2	Weighted means of nine Surber or core samples collected every 2 mo for 1 y on four habitat types.	Stone and Wallace (1998)
1995	1–5 cm	310 (83)	362 (75)	All wood measured in 1-m transects along the stream. Mean (standard error) of 60 transects.	Webster and Golladay (unpublished)
	5–20 cm	1,983 (647)	1,258 (435)		
	>20 cm	13,198 (1,396)	3,209 (1,362)		
1995	<1 cm	51 (6)	—	Line intercept method with lines across the width of the stream. Mean (standard error) of 60 lines. Volume converted to mass using measured densities of 0.41 (BHB) and 0.46 g/cm^3 (HWC).	Webster and Golladay (unpublished)
	1–5 cm	364 (49)	—		
	5–20 cm	2,188 (514)	—		
	>20 cm	14,771 (4,841))	—		
1999	5–20 cm	3,680 (677)	2,251 (342)	Line intercept method, same as 1995	Webster and Golladay (unpublished)
	>20 cm	6,680 (1,990)	4,187 (1,612)		
2004–05	<1 cm	54.0	38.8	Means of 30 core samples collected seasonally at ten transects along each stream	Webster and Benfield (unpublished)
2008	1–5 cm	404	431	Line intercept method along 710 m of BHB and 200 m of HWC. Volume converted to density using 0.4 g/cm^3	Webster and Benfield (unpublished)
	5–20 cm	1,294	1,792		
	>20 cm	1,982	2,559		

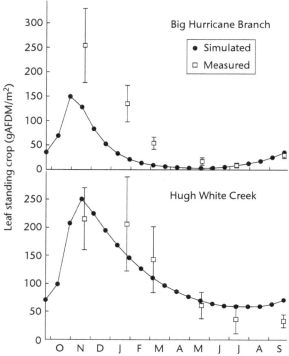

Figure 10.8 Simulation of benthic leaf material in Big Hurricane Branch and Hugh White Creek with just leaf-fall and simple leaf breakdown rates in a four compartment model. Measured data are from 1993–94.

rates of leaf disappearance from mesh bags agree with the loss of leaf mass for the stream. However, this was not the case for Big Hurricane Branch—our simulations were consistently lower than measured standing crop (figures 10.8–10). It appears that leaves disappear from the mesh bags considerably faster than leaves actually disappear from this stream.

Discussion

Our studies indicate that except for the first few years after logging, leaves that fall into Big Hurricane Branch disappear faster than leaves falling into Hugh White Creek. This conclusion is based on both leaf breakdown rates (figure 10.6) and disappearance of benthic leaves (figure 10.7). There are several possible reasons for this difference. First, more leaves may simply wash out of Big Hurricane Branch during storms. Based on measurements made in 1977–78 (Gurtz et al. 1980) and again in 1984–85 (Golladay et al. 1987), particulate organic matter transport

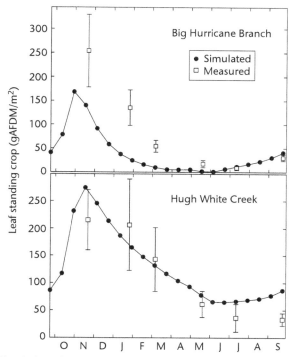

Figure 10.9   Simulation of benthic leaf material in Big Hurricane Branch and Hugh White Creek including leaf-fall, leaf blow-in, and simple leaf breakdown rates in a four compartment model. Measured data are from 1993–94.

from Big Hurricane Branch has been significantly greater than from Hugh White Creek. However, whole leaves and leaf fragments are a small fraction of particulate transport (Gurtz et al. 1980) and cannot account for the observed differences in leaf-disappearance rates.

A factor that clearly did contribute to the differences in disappearance rates was the more labile composition of the litterfall to Big Hurricane Branch for the first 15–20 years following logging (table 10.2). However, more recent data suggest that the leaf species composition falling in the stream is converging toward the characteristics of reference forests. Yet this does not explain the faster species-specific leaf breakdown rates in Big Hurricane Branch.

Temperature is a major factor affecting leaf breakdown rates (Webster and Benfield 1986), and the warmer winter temperature in Big Hurricane Branch (figure 10.3) may contribute to faster leaf disappearance. However, when we take temperature into account by using degree-day adjusted breakdown rates, leaf breakdown rates in Big Hurricane Branch are still much higher than in Hugh White Creek (table 10.3).

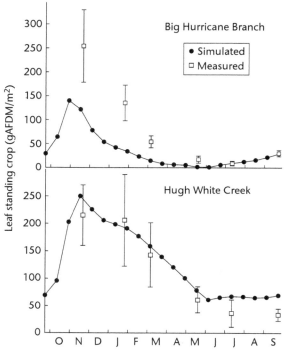

Figure 10.10   Simulation of benthic leaf material in Big Hurricane Branch and Hugh White Creek including leaf-fall, leaf blow-in, and degree-day leaf breakdown rates in a four compartment model. Measured data are from 1993–94.

Heterotrophic processes in Coweeta streams appear to be limited by availability of both nitrogen and phosphorus (Tank and Webster 1998; Gulis and Suberkropp 2003; Greenwood et al. 2007). Dissolved phosphorus levels in Big Hurricane Branch have been very low and apparently little affected by logging (Swank 1988). However, dissolved inorganic nitrogen increased in Big Hurricane Branch following logging, decreased for about 10 years, and then returned to relatively high levels (Swank and Vose 1997) and remains elevated (see Qualls et al., chapter 5, this volume). The availability of inorganic nitrogen is likely a contributing factor to the faster disappearance of leaves in Big Hurricane Branch. This appears to be especially true for rhododendron, which is the most refractory leaf species we studied. Experimental addition of nutrients to a small stream at Coweeta had the same effect—rhododendron leaf breakdown was accelerated more than breakdown of relatively labile red maple leaves (Gulis and Suberkropp 2003; Greenwood et al. 2007).

Physical scouring of leaf packs by transported sediment may also accelerate leaf breakdown in Big Hurricane Branch. Sediment movement and its effect on leaf breakdown may become more evident as large wood continues to decline.

These various factors, quality of leaf inputs, higher winter temperature, higher nutrient availability, and sediment scouring together can probably account for the differences in the rates leaves disappear from the two streams. However, they do not explain why measured leaf breakdown rates in Big Hurricane Branch were so much faster than actual leaf disappearance, whereas these rates were similar in Hugh White Creek (figures 10.8–10). One possibility is that the scarcity of leaves in Big Hurricane Branch in late winter and throughout spring and summer (figure 10.7), results in introduced bags of leaves being a rare resource that was rapidly colonized and broken down by shredding invertebrates (Webster and Waide 1982). Stout et al. (1993) found higher shredder biomass and production in Big Hurricane Branch than in Hugh White Creek in 1986–87, and Stone and Wallace (1998) found a similar pattern in 1993–94. Leaf-shredding invertebrates in Big Hurricane Branch may have responded to the high availability of more labile leaf material in late fall and early winter, and then, in late winter and spring, they congregated in our packs of more refractory litter. As a result, our measured leaf breakdown rates, especially for the more refractory rhododendron, were exceptionally high.

There has been insufficient time since logging to evaluate the prediction that a long-term decrease in large wood will result accelerated sediment movement from Big Hurricane Branch (Webster et al. 1992). Large, slow-decaying logs remain in the stream and undoubtedly contribute to the physical stability of the streambed. In other aspects, 30 years after its watershed was logged, Big Hurricane Branch still shows many effects of the disturbance. Though the stream rapidly returned to a dependence on allochthonous detrital inputs, the continued altered quality of these inputs, coupled with elevated available nitrogen, results in substantially altered detrital dynamics within the stream. Thus Big Hurricane Branch exhibits some evidence of high resilience, but in other aspects potential resilience is still limited by the press-type nature of forest clearing. Until the riparian forest returns to its predisturbance state, we expect Big Hurricane Branch to show signs of altered ecological structure and function. Even with eventual forest recovery, we predict that stream recovery will show a significant temporal lag. In reality, with the permanent forest changes that have occurred in the region, such as the loss of chestnut and the alteration of atmospheric nutrient inputs as well as future changes, such as loss of hemlocks and climate change, we predict that no stream within the southern Appalachians will return to presettlement conditions. The best we can hope for in recently or intensively disturbed streams, like Big Hurricane Branch, is a return to a state that that contains many of the original species of the stream community and regains most of its functional characteristics.

Acknowledgments

This study was supported by grants from the National Science Foundation, since 1982 as part of the Long-Term Ecological Research Program.

Literature Cited

Bender, E. A., T. J. Case, and M. E. Gilpin. 1984. Perturbation experiments in community ecology: theory and practice. *Ecology* 65: 1–13.

Benfield, E. F. 2006. Leaf breakdown in stream ecosystems. Pages 711–720 in *Methods in Stream Ecology*. F. R. Hauer and G. A. Lamberti, editors. Academic Press, San Diego, California.

Benfield, E. F., J. R. Webster, S. W. Golladay, G. T. Peters, and B. M. Stout. 1991. Effects of forest disturbance on leaf breakdown in southern Appalachian streams. *Verhandlung der Internationalen Vereinigung für Theoretische und Angewandte Limnologie* 24: 1687–1690.

Benfield, E. F., J. R. Webster, J. L. Tank, and J. J. Hutchens. 2001. Long-term patterns in leaf breakdown in streams in response to watershed logging. *International Review of Hydrobiology* 86: 467–474.

Boring, L. R., C. D. Monk, and W. T. Swank. 1981. Early regeneration of a clear-cut southern Appalachian forest. *Ecology* 62: 1244–1253.

Golladay, S. W., and J. R. Webster. 1988. Effects of clearcut logging on wood breakdown in Appalachian mountain streams. *American Midland Naturalist* 119: 143–155.

Golladay, S. W., J. R. Webster, and E. F. Benfield. 1987. Changes in stream morphology and storm transport of organic matter following watershed disturbance. *Journal of the North American Benthological Society* 6: 1–11.

Golladay, S. W., J. R. Webster, and E. F. Benfield. 1989. Changes in stream benthic organic matter following watershed disturbance. *Holarctic Ecology.* 12: 96–105.

Greenwood, J. L., A. D. Rosemond, J. B. Wallace, W. F. Cross, and H. S. Weyers. 2007. Nutrients stimulate leaf breakdown rates and detritivore biomass: bottom-up effects via heterotrophic pathways. *Oecologia* 151: 637–649.

Gulis, V., and K. Suberkropp. 2003. Leaf litter decomposition and microbial activity in nutrient-enriched and unaltered reaches of a headwater stream. *Freshwater Biology* 48: 123–134.

Gurtz, M. E. 1981. Ecology of stream invertebrates in a forested and a commercially clear-cut watershed. PhD thesis. University of Georgia; Athens.

Gurtz, M. E., and J. B. Wallace. 1984. Substrate-mediated response of stream invertebrates to disturbance. *Ecology* 65: 1556–1569.

Gurtz, M. E., J. R. Webster, and J. B. Wallace. 1980. Seston dynamics in southern Appalachian streams: effects of clear-cutting. *Canadian Journal of Fisheries and Aquatic Sciences* 37: 624–631.

Hynes, H. B. N. 1975. The stream and its valley. *Verhandlung der Internationalen Vereinigung für Theoretische und Angewandte Limnologie* 19: 1–15.

Minshall, G. W., R. C. Petersen, K. W. Cummins, T. L. Bott, J. R. Sedell, C. E. Cushing, and R. L. Vannote. 1983. Interbiome comparison of stream ecosystem dynamics. *Ecological Monographs* 53: 1–25.

Mulholland, P. J., E. R. Marzoff, J. R. Webster, D. R. Hart, and S. P. Hendricks. 1997. Evidence that hyporheic zone increase heterotrophic metabolism and phosphorus uptake in forest streams. *Limnology and Oceanography* 42: 443–451.

Rosenzweig, M. L., and R. H. MacArthur. 1963. Graphical representation and stability conditions of predator-prey interactions. *American Naturalist* 97: 209–223.

Stone, M. K., and J. B. Wallace. 1998. Long-term recovery of a mountain stream from clear-cut logging: the effects of forest succession on benthic invertebrate community structure. *Freshwater Biology* 39: 151–169.

Stout, B. M., E. F. Benfield, and J. R. Webster. 1993. Effects of a forest disturbance on shredder production in southern Appalachian headwater streams. *Freshwater Biology* 29: 59–69.

Swank, W. T. 1988. Stream chemistry responses to disturbance. Pages 339–357 in *Forest Ecology and Hydrology at Coweeta*. W. T. Swank and D. A. Crossley, Jr., editors. Springer-Verlag, New York, New York.

Swank, W. T., and J. M. Vose. 1997. Long-term nitrogen dynamics of Coweeta forested watersheds in the southeastern United States of America. *Global Biogeochemical Cycles* 11: 657–671.

Swift, L. W., Jr. 1983. Duration of stream temperature increases following forest cutting in the southern Appalachian mountains. Pages 273–275 in A. I. Johnson and R. A. Clark, editors. International Symposium on Hydrometeorology. American Water Resources Association, Bethesda, Maryland.

Tank, J. T., and J. R. Webster. 1998. Interaction of substrate and nutrient availability on wood biofilm processes in streams. *Ecology* 79: 2168–2179.

Vannote, R. L., G. W. Minshall, K. W. Cummins, J. R. Sedell, and C. E. Cushing. 1980. The river continuum concept. *Canadian Journal of Fisheries and Aquatic Sciences* 37: 130–137.

Waide, J. B., and J. R. Webster. 1976. Engineering systems analysis: applicability to ecosystems. Pages 329–371 in *Systems Analysis and Simulation in Ecology*. B. C. Patten, editor. Academic Press, New York, New York.

Wallace, J. B., and A. C. Benke. 1984. Quantification of wood habitat in subtropical coastal plain streams. *Canadian Journal of Fisheries and Aquatic Sciences* 41: 1643–1652.

Wallace, J. B., J. R. Webster, S. L. Eggert, J. L. Meyer, and E. R. Siler. 2001. Large woody debris in a headwater stream: long-term legacies of forest disturbance. *International Review of Hydrobiology* 86: 501–513.

Warren, W. G., and P. F. Olsen. 1964. A line intersect technique for assessing logging waste. *Forest Science* 10: 267–276.

Webster, J. R., and E. F. Benfield. 1986. Vascular plant breakdown in freshwater ecosystems. *Annual Review of Ecology and Systematics* 17: 567–594.

Webster, J. R., E. F. Benfield, T. P. Ehrman, M. A. Schaeffer, J. L. Tank, J. J. Hutchens, and D. J. D'Angelo. 1999. What happens to allochthonous material that falls into streams? A synthesis of new and published information from Coweeta. *Freshwater Biology* 41: 687–705.

Webster, J. R., E. F. Benfield, J. J. Hutchens, J. L. Tank, S. W. Golladay, and J. C. Adams. 2001. Do leaf breakdown rates actually measure leaf disappearance from streams? *International Review of Hydrobiology* 86: 417–427.

Webster, J. R., S. W. Golladay, E. F. Benfield, D. J. D'Angelo, and G. T. Peters. 1990. Effects of forest disturbance on particulate organic matter budgets of small streams. *Journal of the North American Benthological Society* 9: 120–140.

Webster, J. R., S. W. Golladay, E. F. Benfield, J. L. Meyer, W. T. Swank, and J. B. Wallace. 1992. Catchment disturbance and stream response: an overview of stream research at Coweeta Hydrologic Laboratory. Pages 231–253 in *River Conservation and Management*. P. J. Boon, P. Calow, and G. E. Petts, editors. Wiley, Chichester, England.

Webster, J. R., M. E. Gurtz, J. J. Hains, J. L. Meyer, W. T. Swank, J. B. Waide, and J. B. Wallace. 1983. Stability of stream ecosystems. Pages 355–395 in *Stream Ecology*. J. R. Barnes and G. W. Minshall, editors. Plenum Press, New York, New York.

Webster, J. R., J. L. Meyer, J. B. Wallace, and E. F. Benfield. 1997. Organic matter dynamics in Hugh White Creek, Coweeta Hydrologic Laboratory, North Carolina, USA.

Pages 74–78 in J. R. Webster and J. L. Meyer, editors. *Stream organic matter budgets.* Special issue of *Journal of the North American Benthological Society* 16: 3–161.

Webster, J. R., K. Morkeski, C. A. Wojculewski, B. R. Niederlehner, E. F. Benfield, and K. J. Elliott. 2012. Effects of hemlock mortality on streams in the southern Appalachian Mountains. *American Midland Naturalist* 168: 112–131.

Webster, J. R., and B. C. Patten. 1979. Effects of watershed perturbation on stream potassium and calcium dynamics. *Ecological Monographs* 49: 51–72.

Webster, J. R., and J. B. Waide. 1982. Effects of forest clearcutting on leaf breakdown in a southern Appalachian stream. *Freshwater Biology* 12: 331–344.

Webster, J. R., J. B. Waide, and B. C. Patten. 1975. Nutrient recycling and the stability of ecosystems. Pages 1–27 in F. G. Howell, J. B. Gentry, and M. H. Smith, editors. Mineral cycling in Southeastern ecosystems. ERDA Symposium Series. CONF-740513. National Technical, Information Service. Springfield, Virginia.

11

Stream Macroinvertebrate Response to Clearcut Logging

J. Bruce Wallace*
Damon Ely

Introduction

Why study response of stream invertebrates to watershed disturbances such as clearcut logging? Stream invertebrates can be excellent integrators of changes in such ecosystem phenomena as changes in the food base of ecosystems. For example, a number of invertebrate taxa appear to track changes in food resources (Hawkins et al. 1982; Fuller et al. 1986; Wallace and Gurtz 1986; Feminella and Hawkins 1995; Wallace and Webster 1996; Wallace et al. 1997; Jackson et al. 2007). Many taxa also exhibit substrate-specific as well as taxon-specific responses to physical alterations, such as sediment addition (Gurtz and Wallace 1984; Zweig and Rabeni 2001). Benthic invertebrates can also play important roles in many ecological processes (Wallace and Webster 1996), and in whole ecosystems, benthic invertebrates often display early and dramatic responses to manipulation (Reice and Wohlenberg 1993). Thus, assessment of benthic assemblages can be a valuable means to detect disturbance and long-term recovery from disturbance. Furthermore, knowledge of a change in the community structure of invertebrates offers an additional mechanism for detecting changes in ecological processes through time. Unfortunately, there have been few studies of the response to clearcut logging within the same stream over many years. Rather, such changes are occasionally inferred from examining multiple streams draining catchments that are in different stages of recovery.

Overview of Invertebrate Studies

Invertebrates in Big Hurricane Branch (BHB) draining Watershed (WS) 7 (clearcut) and Hugh White Creek (HWC) draining Watershed (WS) 14 (reference) were examined several times starting in 1977 immediately prior to the clearcut and following logging. During the original clearcut period, benthic samples were taken

* Corresponding author: Department of Entomology, University of Georgia, Athens, Georgia 30602 USA

monthly from January 1977 through September 1978 (Gurtz and Wallace 1984). Three logging roads for access were constructed on the watershed between April and June 1976 and logging encompassed the period of January 1977 to June 1977 (see Swank and Webster, chapter 1, this volume). Most of the logging debris entering the stream channel was removed between August and October 1977. In 1982–83, five years following the initial logging, a series of seasonal (every 3 months) benthic samples were taken in each stream, BHB and HWC (Wallace et al. 1988). Between February 1993 and February 1994 both BHB and HWC were sampled bimonthly (every 2 months) for benthic invertebrates (Stone and Wallace 1998). Benthic samples were again collected bimonthly between February 2003 and December 2003 (Ely 2005; Ely and Wallace 2010). In the 1977–78 study, 4 samples were collected in bedrock outcrop, cobble riffle, pebble riffle, and sandy reach (depositional) areas of each stream reach (upper, middle, and lower) using a stratified random sampling procedure (Gurtz and Wallace 1984). In the 1982–83 and 1993–94 collecting periods, a similar stratified-random sampling procedure was used in which nine samples were collected from the upper, middle, and lower portion of each stream (Wallace et al. 1988). However, in these latter studies we limited substrate sampling to bedrock outcrop, cobble-pebble riffles, and depositional areas, that is, 3 substratum types. The 2003 study employed a similar sampling regime to the 1993–94 study; however, the bedrock outcrop habitat was further divided into areas with high moss cover and with low moss cover. During the original 1977–78 and 1982–83 studies, a Surber sampler was used for all collections. For the later studies, 225-cm^2 areas of bedrock were scraped into a 250-µm mesh net; a Surber was used for cobble-riffle samples; and a 400-cm^2 stove pipe corer was used to sample depositional habitats. Between-year comparisons of benthic abundances were complicated for depositional habitats because the benthic corer greatly increased the recovery of small animals, especially collector-gatherer taxa.

For the 1993–94 and 2003 studies, we examined abundance, biomass, and secondary production for most taxa (Stone and Wallace 1998; Ely 2005), whereas the earlier studies had focused primarily on abundance and biomass. However, for some selected taxa in the 1977–78 study that showed especially strong responses to clearcutting, we also measured production (Gurtz and Wallace 1986; Wallace and Gurtz 1986). Production is the most comprehensive measure of the success of a population because it includes a composite of several features: abundance, biomass, growth, reproduction, survivorship, and generation time (Benke 1993). Productivity is most useful when assessing the impact of animals on their resources and energy flow through populations. An example is given below under the baetid mayfly responses in BHB to clearcutting and subsequent canopy regrowth of WS 7.

Results

Changes in Physical Properties and Resource Base

Stream invertebrates represent an important response variable to watershed disturbances, such as logging, because they respond both directly and indirectly to many of the physical and biological disturbances associated with clearcut timber harvest. One of the most striking changes was increased solar radiation and within-stream primary production. In the early clearcutting stages of WS 7, there were reduced allochthonous litter inputs and a short-lived shift from an allochthonous to an autochthonous energy base (Webster et al. 1983; Wallace and Gurtz 1986; Wallace 1987; for a detailed account of changes in organic matter standing crop in BHB and HWC, see Webster et al., chapter 10, this volume). Dissolved organic matter and nutrient inputs from the terrestrial ecosystem were altered as a result of vegetation removal (Meyer and Tate 1983; Swank and Waide 1988). Stream thermal regimes were altered as a result of the removal of terrestrial vegetation (Swift 1983). Sediment inputs were increased as a result of road-building activities and increased soil disturbance (Gurtz et. al. 1980; Webster et al. 1983; Gurtz and Wallace 1984). Increased runoff resulted from loss of terrestrial vegetation and lower evapotranspiration (Swank et al. 1982). With such a wide array of disturbances to both the energy base and physical environment, it is not surprising that stream invertebrate populations would exhibit strong responses to logging.

Invertebrate Response

Total benthic abundance of invertebrates increased in BHB as compared to HWC following logging, and most of this increase occurred in the moss-covered bedrock substrates (Gurtz and Wallace 1984). Much of this increase in abundance was attributable to the collector-gatherer (C/G) functional feeding group, especially baetid and *Ephemerella* spp. mayflies (Gurtz and Wallace 1984). However, subsequent work and gut analyses of the baetid mayflies showed large amounts of diatoms in their guts, suggesting that these animals were functioning more as scrapers than collector-gatherers in these two Coweeta streams (Wallace and Gurtz 1986). The failure of a significant habitat-weighted scraper response to the clearcutting of WS 7 reported by Gurtz and Wallace (1984) is undoubtedly attributable to the placement of baetid mayflies in the collector functional group rather than the scraper group.

We initially hypothesized that the shredder functional group would decline following clearcutting because of reduced litter inputs. Large reductions in litter inputs did occur as leaf-fall inputs to BHB declined from 259.2 g m^{-2} y^{-1} prior to logging to 4.2 g in 1978 and 1979 following logging, and annual lateral movement inputs declined from 178 g/linear m prior to logging to 38.6 g following logging (Webster and Waide 1982). Despite these large reductions in litter inputs to BHB, abundance of the shredder functional feeding group did not decline as expected (Gurtz and Wallace 1984). Although there was a large reduction in leaf litter inputs, no significant difference was found in leaf detritus in the cobble riffle habitats between BHB and HWC during the 21-month post-logging sampling period (Gurtz and Wallace 1984). Standing crop of leaf detritus only declined in sandy and pebble substrates of BHB, and much of the leaf material was buried in the substrate or retained in dams of woody debris. For a nearby Coweeta stream, where all terrestrial litter inputs were excluded from a headwater stream, it took 2 years to see significant reductions in shredder abundance and biomass (Wallace et al. 1997, 1999). Thus the 21-month post-logging sampling period, coupled with inputs from logging and material retained in the channel may have been insufficient to produce a detectable "bottleneck" in shredder food resources. In other streams, increases in shredder taxa immediately following logging may occur when logging slash is not removed and detrital resources increase (Jackson et al. 2007). Reduced litterfall inputs to BHB did not persist for a long period due to the rather rapid plant succession on WS 7. For example, within 6 years following clearcutting, litter inputs to BHB were about 89% (436 g AFDM m^{-2} yr^{-1}) of that of the reference stream HWC (491 g AFDM m^{-2} yr^{-1}). However, the standing crop of large (> 1 mm diameter) benthic organic matter in BHB was only 58% of that of HWC; and fine organic matter (< 1 mm diameter) in BHB was 68% of HWC (Golladay et al. 1989). In a study on small tributary streams within WS 7 and WS 14, completed 9 to 10 years following clearcutting (1986–1987), litter inputs to BHB contained more fast-processing leaf species than those to HWC. Abundance, biomass, and production of the three dominant shredder taxa, *Tallaperla* (Plecoptera), *Pycnopsyche* (Trichoptera), and *Tipula* (Diptera), were all higher in the disturbed tributaries of BHB than in the reference tributaries of HWC (Stout et al. 1993). Stout et al. (1993) attributed the increased abundance, biomass, and production of these shredders to greater inputs of early successional species of litter, which were 7.5 times higher in BHB tributaries than in those of HWC. Production of shredder taxa was also 3.8 times higher in the main channel of BHB than in HWC 17 years following the clearcut (Stone and Wallace 1998). Twenty-six years after logging, shredder production was similar between streams; however, the annual standing crop of benthic detritus in BHB was only 49% of that in HWC (Ely 2005).

As a group, collector-gatherers increased on all substrates except sand in BHB following logging as compared with the reference stream HWC. However, with exception of collector-gatherer chironomids (Diptera) on bedrock substrates and *Ephemerella* spp. mayflies on all substrates, much of the increase was attributable to baetid mayflies, which we now know were functioning more like scrapers than collector-gatherers (Wallace and Gurtz 1986). In 1993–94, production of collector-gatherers remained greater in all habitats of BHB than of HWC. These differences were most pronounced in the moss-covered bedrock habitats of BHB, where abundance exceeded HWC by 12 times and production of collector-gatherers exceeded those of HWC by 4.3 times (Stone and Wallace 1998). In 2003, abundance, biomass, and production of collector-gatherers were similar in all habitats of BHB and HWC except bedrock habitats with high moss cover, where all values were greater in BHB by 2- to 3-fold (Ely 2005).

The response of baetid mayflies to the clearcut were so striking, we measured abundance, biomass, production, and gut analyses of these mayflies in both streams (Wallace and Gurtz 1986). Mean standing stock abundance and biomass of baetids increased in all substrate types (bedrock, cobble riffle, pebble riffle, and sandy reach) of BHB compared to HWC, and baetids reached their peak in BHB about 12 to 21 months following logging (Wallace and Gurtz 1986). Habitat-weighted baetid production in BHB averaged 17 times that of the reference stream, HWC. Baetid production increased during logging and increased again following site preparation when logging slash was removed from the stream bed and riparian rhododendron was cut (figure 11.1). Abundance, biomass, and production were much greater in

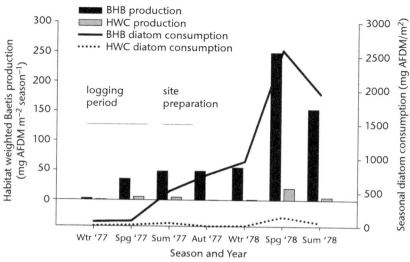

Figure 11.1   Habitat-weighted baetid mayfly production in Hugh White Creek (HWC) and Big Hurricane Branch (BHB) during successive seasons starting with the logging period of BHB (HWC was the reference stream). Note the large increase in baetid production following site preparation (removal of slash and riparian rhododendron) in BHB. Much of the increase in *Baetis* production was attributable to diatom consumption. *Baetis* consumption of diatoms was estimated based on the methods of Benke and Wallace (1980). Redrawn from the data of Wallace and Gurtz (1986).

BHB than HWC, and following site preparation, baetid guts examined from BHB contained more diatoms than those from HWC (figure 11.2). Increased diatoms in guts, as well as increased abundance and biomass of baetids in BHB, indicates a much greater rate of algal grazing by baetids in the clearcut stream. The large increased abundance and production of baetids and other scraper taxa are probably why Hains (1981) could not detect an increase in algal cells in BHB compared to HWC, despite measuring higher rates of primary productivity in BHB. In the more physically stable bedrock habitat, baetid production in BHB was almost 28 times that of HWC (Wallace and Gurtz 1986). Surprisingly, diatoms comprised the most important component of the baetid diets in each stream, even the heavily shaded reference stream. Estimated diatom consumption to account for baetid production was 25 times greater in BHB than HWC (5.788 versus 0.234 g AFDM/m2) (figure 11.2). Baetids obviously responded to changes in the energy base of headwater streams, that is, the shift from allochthonous energy resources to autochthonous resources immediately following logging (Hains 1981). However, the increase in primary productivity was short-lived (Webster et al. 1983), and baetid abundance continued to decline in 1982–83 (Wallace et al. 1988) and 1992–93 (Stone and Wallace 1998) with increased shading and terrestrial litter inputs. The high baetid production in BHB, 17 times that of HWC, observed in the first 21 months following logging, had declined to 3.4 times that of HWC by 1992–93 (Stone and Wallace

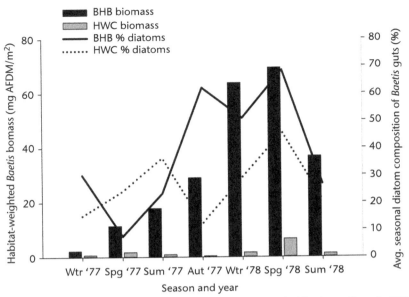

Figure 11.2   Seasonal habitat-weighted baetid biomass in Big Hurricane Branch (BHB), which drains the clearcut watershed, and Hugh White Creek (HWC), which drains the reference stream. The logging and site preparation times for BHB are noted in figure 11.1. Note that following site preparation, baetid populations in BHB contained more diatoms in their guts, despite much higher biomass (and abundance) than in the reference stream HWC, indicating much greater consumption of diatoms in the clearcut stream. Percent diatoms in baetid guts based on projected area of gut contents. Data are from Wallace and Gurtz (1986).

1998). By 2003, mean annual habitat-weighted baetid production was nearly equal between streams (Ely 2005).

Immediately following logging, filter-feeding invertebrates declined significantly in the depositional substrates of BHB compared to the reference stream, HWC. Hydropsychid caddisflies dominated abundance and production of filterers in both BHB and HWC and displayed distinct substrate preferences with bedrock outcrops > riffles > depositional areas (Gurtz and Wallace 1986). The same substrate-specific trends for filterers existed 17 years (Stone and Wallace 1998) and 26 years later (Ely 2005) in each stream.

During the initial clearcut, the abundance of invertebrate predators on bedrock substrates increased significantly in BHB and decreased significantly in depositional areas compared to HWC (Gurtz and Wallace 1984). In 1993–94, invertebrate predator production in BHB exceeded that of HWC for all habitats and was greatest for the bedrock outcrop habitats; that is, bedrock outcrop (6.3 x) > riffle (1.7 x) > depositional (1.4 x) (Stone and Wallace 1998). Undoubtedly, the greater predator production in BHB was related to the approximately 2 times greater overall production in BHB than was found in HWC during that study. Other studies at Coweeta have found evidence that production of invertebrate predators is closely linked to prey production (Wallace et al. 1997, 1999). Thus the greater abundance of prey in BHB may have reduced food limitation on predatory invertebrates. In 2003, predator production in BHB was only 1.2 times that in HWC, reflecting the similarity in whole-assemblage production between BHB (10.1 g AFDM m^{-2} yr^{-1}) and HWC (9.3 g AFDM m^{-2} yr^{-1}) during that study (Ely 2005).

Synthesis

Benthic Biotic Indices Following Logging

Disturbance may be considered to be the movement of the community away from a nominal value or behavior (Yount and Niemi 1990); and recovery, as directional movement toward some state resembling the predisturbance state. Biological indices are often used to evaluate the impact of disturbance on stream communities. The choice of indices, sampling methods, sampling locations, and reference streams may all affect the ability to determine the level of recovery (Niemi et al.1993). Preferably, indices should show a definitive trend between reference and disturbed streams compared to initial conditions. There should also be relatively little variation of the index in the reference stream over the same period (Fortino et al. 2004).

Taxonomic diversity usually decreases with disturbance and increases with recovery. However, based on observations from BHB, clearcut harvesting may result in a somewhat different scenario. Immediately after clearcutting, EPT (Ephemeroptera, Plecoptera, and Trichoptera) taxa increased in the disturbed stream compared with the reference. Increased insolation, primary productivity, and altered thermal regimes (Swift, 1983) occurred in the disturbed stream, which made it more similar to downstream reaches. Several taxa normally confined to larger downstream reaches, for example, *Pteronarcys* (Plecoptera) and *Hydropsyche* (Trichoptera), colonized the disturbed stream without significant loss of headwater EPT taxa (Gurtz 1981). Following catchment logging of a low-gradient Virginia stream, more invertebrate taxa were also found in a logged reach versus a nearby forested reach (Kedzierski and Smock 2001).

Clearcutting caused shifts in the predominance of certain taxa in the disturbed stream, but few taxa were lost from the community. Habitat-weighted EPT indices of taxa richness and abundance show subsequent declines in BHB as forest succession, that is, recovery, progressed. Taxa richness may thus be more appropriate for detecting such disturbances as organic pollution and toxic chemicals, which result in the loss of sensitive species, rather than disturbances that only cause shifts in relative abundance. However, Martel et al. (2007) found declining richness in larger streams ($\geq$ third order) when basin-wide logging intensity increased, suggesting that cumulative impacts of logging may result in different invertebrate responses in downstream reaches than in smaller headwaters and thus require alternative biomonitoring techniques.

Among the indices used by Stone and Wallace (1998), the percentage *Baetis* index, shredder-scraper index, and the North Carolina Biotic Index (NCBI; Lenat 1993) showed the greatest ability to detect differences between the clearcut and reference streams through time. During the initial study, the percentage *Baetis* index indicated differences between reference and disturbed streams in all habitats followed by decreases in all habitats during subsequent years. Others have also noted a large increase in *Baetis* in streams draining deforested catchments (Newbold et al. 1980; Noel et al. 1986; Anderson 1992). There was a trend of decreasing percentage *Baetis* index in the disturbed stream during the 1982, 1993, and 2003 studies, while the reference stream was relatively constant. The trend of decreasing *Baetis* following clearcutting of WS 7 indicated that differences between reference and disturbed streams in 1982, 1993, and 2003 were insignificant. The reduction of *Baetis* populations was correlated with increased shading of the stream by riparian vegetation during succession. The percentage *Baetis* index measured was one of the major long-term trends identified by Stone and Wallace (1998), along with decreasing abundance of scrapers in response to forest succession and the return to an allochthonous energy base.

The shredder-scraper index also showed a significant difference between reference and disturbed streams only during the first five years following clearcutting. The index was extremely low in the disturbed stream during 1977, indicating lower shredder and higher scraper abundance, and though still low in 1982, had increased over the 5-year period. In contrast, by 1993 the shredder-scraper index in BHB far surpassed that of HWC, indicating a community dominated by shredders with relatively few scrapers. By 2003, the shredder-scraper ratio in BHB had lowered and was more similar between streams than in any previous year (Ely 2005). The reference stream shredder-scraper index was relatively constant over the 4 study periods, although the depositional habitat did show some variation. This index combines the two distinct trends in macroinvertebrate abundance identified by Stone and Wallace (1998) compared to 1977 and 1982, namely, increasing shredders and decreasing scrapers. Thus, there were ongoing changes in functional structure of the benthic community in the clearcut stream between 1977 and 1993, with apparent recovery by 2003 (Ely 2005).

The relative rate of recovery also differed among habitats with most rapid recovery in the more physically stable habitat, moss-covered bedrock. However, a slower rate of recovery of biotic indices in the riffle and deposition habitats (Stone and Wallace 1998) suggests that these two habitats may be more sensitive indicators of long-term recovery than moss-covered bedrock habitats (Stone and Wallace 1998). The different rate of recovery of biotic indices among habitats also emphasizes the need for sampling regimes that cover all major habitats when assessing rates of recovery.

As pointed out by others (e.g., Karr 1994; Stone and Wallace 1998), biotic indices vary according to the nature and type of disturbance, that is, pulse versus press (sensu Bender et al. 1984). A number of studies have demonstrated the fallacy of relying on only one index for all disturbances (Karr 1994; Stone and Wallace 1996); multimetric monitoring, predictive modeling, or both approaches are recommended (Fortino et al. 2004). Treatment of a nearby Coweeta stream with an insecticide resulted in similar trends for both EPT taxa richness and the NCBI (Wallace et al. 1996). However, the clearcutting of WS 7 resulted in increased EPT taxa richness and the movement of downstream taxa into the upstream clearcut reach, without corresponding loss of upstream taxa (Stone and Wallace 1998). Contrary to taxa richness, the NCBI incorporates both tolerance and relative abundance of taxa and appears to be more sensitive for measuring subtle changes following logging as the average NCBI in the disturbed stream showed gradual decline (lower score = more sensitive species) from 1977 to 2003. During this same 26-year period, the NCBI of the reference stream, HWC, remained virtually unchanged (Stone and Wallace 1998; Ely 2005).

The five indices calculated from data collected over the past 26 years, as well as the abundance, biomass, and production data collected during the two most recent studies, proved to be of differing value in assessing recovery of the disturbed stream from logging. Percent dominant taxon and EPT richness failed to show any initial differences between reference and disturbed streams indicating that these indices may not be useful for measuring recovery from logging. The percentage *Baetis* and shredder-scraper indices showed significant differences only during the 1977 study and suggest some recovery (no difference between reference and disturbed) by 1982. The NCBI showed continued differences during 1982 in the riffle and depositional habitats and recovery by 1993. Comparisons of total macroinvertebrate abundance, biomass and production, and EPT abundance, indicated continued differences between the reference and disturbed streams in 1993 and apparent recovery by 2003 (Ely 2005). Thus, while benthic community composition had recovered by 1993 (i.e., similar NCBI scores between reference and disturbed streams), the much greater production in BHB than HWC implies that functional processes remained different between these streams.

Although total secondary production converged by 2003, these similarities reflect changes in the reference stream and underscore the role of detrital resources in the functioning of these headwater stream ecosystems. Macroinvertebrate production in HWC more than doubled from 1993 to 2003 simultaneous with a doubling of standing crop of detritus (Ely 2005). In BHB, neither secondary production nor detrital standing crops changed substantially during this time (Ely 2005). Positive relationships between secondary production and resource standing crops in low-order streams have been observed at Coweeta (Wallace et al. 1997, 1999) and elsewhere (Grubaugh et al. 1997; Hall et al. 2001). Similarly, the standing crop of benthic organic matter was a strong predictor of macroinvertebrate biomass in both streams during 2003 (Ely 2005). Thus, the BHB assemblage in 2003 elaborated biomass at a similar rate to HWC despite having only half the available detrital resources, indicating continued differences in consumer-resource relationships most likely related to differences in leaf type (i.e., greater inputs of early successional species to BHB) and the ongoing nutrient enrichment of BHB (Swank et al. 2001). However, we caution that a standing crop of food resources does not necessarily measure inputs, fluxes, and utilization of organic matter through the system. These findings suggest that long-term patterns in the input and retention of allochthonous resources following logging are important to the recovery of macroinvertebrate structure and function. Furthermore, these complex consumer-resource relationships may not necessarily be reflected in biotic index scores. In addition to sediment and changes in stream flow, logging can cause major changes in the energy base, nutrient dynamics, light levels, and temperature of streams. In New England, abundances of stream macroinvertebrates in 2-year and 3-year clearcuts were 2–4 times greater than in streams draining uncut reference catchments (Noel et al. 1986). Noel et al. (1986) attributed increased abundance primarily to higher periphyton and stream temperatures in streams draining logged catchments. Similar results were found in British Columbia, where invertebrate abundance and biomass were greater in recently (≤ 5 years) logged boreal streams than in either reference streams or streams draining sites logged 20–25 years earlier, which did not differ (Fuchs et al. 2003). Using multivariate techniques, Fuchs et al. (2003) related the positive invertebrate response in the recently logged streams to higher chlorophyll a biomass and lower canopy cover than the reference and older logged streams. Such shifts in the food base may be accompanied by large increases in scraper secondary production as previously observed for BHB (Wallace and Gurtz 1986 and figures 11.1, 11.2). Increases in secondary production may occur following clearcutting because of increased food quality due to elevated primary production (Webster et al. 1983) and somewhat altered thermal regimes. Furthermore, the associated decrease in leaf detritus immediately following clearcutting may also increase nutrient availability to periphyton because this reduces the competition between leaf-associated microbes and periphyton for limited nutrients (Tank and Webster 1998). During the first 5 years after logging, there was elevated nutrient export from WS 7 (Swank 1988), which would also have enhanced both primary production and microbial activity on detritus. Secondary production can be an ancillary estimate of energy flow through the system and considers differences in turnover rates among organisms (Benke et al.1984). Long-term assessment of secondary production for such groups as shredders and scrapers, rather than abundance measures, would undoubtedly change our assessment of benthic recovery from logging since our results indicate that increased secondary production persists for longer time spans than do changes in major biotic indices in BHB (Stone and Wallace 1998). However, higher levels of secondary production should not necessarily be equated with improved biological conditions. For example, elevated nutrients (Krueger and Waters 1983; Cross et al. 2006, 2007) and severe organic pollution may result in the exceptionally high productivity of a few taxa (Benke 1993). While production is a valuable tool for assessing changes in energy flow and assimilation in the benthic fauna, the expense and time associated with production studies are probably unfeasible for many biomonitoring programs.

Forest-Stream Linkages and Clearcutting

Linkages between stream invertebrates and the surrounding forest include both physical and biotic influences. For example, the forest shades the stream, thus modulating stream temperature (Ross 1963), reducing light availability and within-stream autochthonous production, and providing the allochthonous organic detritus that fuels the microbial community and energy base of the invertebrate community (Wallace et al. 1997, 1999). In contrast, clearcutting reduces allochthonous inputs and increases solar inputs and primary production of streams. The invertebrate community responds to the large-scale shifts in the energy base of the stream following clearcutting accordingly (Haefner and Wallace 1981; Gurtz and Wallace 1984; Wallace and Gurtz 1986). The results of these Coweeta studies emphasize the strong linkages between the forest and headwater streams. The change in energy base immediately following the clearcutting of WS 7 and the subsequent shift in composition of invertebrate assemblages toward short-lived taxa, such as grazing baetid mayflies, demonstrates the close linkage between the surrounding forest and the stream. Elevated summer temperatures, with increased solar input, coupled with increased primary production (Hains 1981) resulted in a headwater stream more typical of downstream reaches. Several taxa that are more typical of downstream areas, such as baetids, *Ephemerella* spp. mayflies, and *Hydropsyche* spp. caddisflies, increased in BHB immediately following clearcutting. Interestingly, Edington and Hildrew (1973) noted that *Hydropsyche* spp. from downstream reaches appeared in a United Kingdom headwater stream following clearcutting; whereas the only hydrodpsychid genus in their study stream prior to logging was *Diplectrona*. Edington and Hildrew's research strongly suggests that different bioenergetic efficiencies in response to changing stream temperatures influenced the ability of *Hydropsyche* to persist within the cooler headwater reaches.

Furthermore, we suspect that our estimates of baetid production are extremely conservative, since differences in growth rates in BHB and HWC were based only on growth rates at various temperatures taken from the European literature and did not take in account differences in food quality or quantity (Wallace and Gurtz 1986). Subsequent studies on the Ogeechee River in southeastern Georgia (Benke and Jacobi 1994) and even in tundra streams in Alaska (A. D. Huryn, personal communication) suggest considerably higher growth rates than we used for BHB and HWC. In future studies, it would be desirable to use field growth chambers (e.g., Huryn and Wallace 1986) to obtain growth rates for selected invertebrates in disturbed and undisturbed streams.

Based on the studies of forest disturbance and aquatic invertebrate productivity from the Coweeta manipulations, it appears that in most cases forest disturbance increases overall invertebrate productivity (Stone and Wallace 1998; Kedzierski and Smock 2001). However, few studies have actually examined the productivity of benthic assemblages following logging but instead infer higher production rates from increased abundance (Nislow and Lowe 2006), biomass (e.g., Fuchs et al. 2003), or increased emergence rates (Banks et al. 2007). Furthermore, productivity should not always be considered the best measure of response to disturbance. For example, some of the highest measures of lotic invertebrate production are associated with excessive organic pollution, such as the heavy organic enrichment of a castle mote in Wales (Benke 1993). Likewise, higher biological diversity may not necessarily be a measure of enhanced habitat quality if there has been invasion by exotic species, such as the influx of downstream species into the headwaters of BHB, as a result of altered physical attributes and energy inputs. Exotic species can induce large changes in aquatic communities (e.g., Huryn 1998). Conservation should strive to preserve the original inhabitants of a given habitat rather than focus on taxa diversity and productivity.

Conclusions

This clearcut was not "typical" as required by recent regulations in most regions of the United States, as there was no riparian zone set aside. Even riparian *Rhododendron* spp., which provide year-round shading of the stream, was cut following logging. If a riparian management zone had been established, the invertebrate response would probably be less pronounced than was observed from several aspects. Unforeseen natural events also increased the magnitude of the study. For example, shortly following road building on WS 7, an exceptionally large storm in the last two weeks of May 1976 resulted in large amounts of sediment entering BHB. This sediment input was greatest where access roads crossed BHB in its middle and upper portions. There was a large increase in inorganic and organic seston in BHB compared to the reference stream HWC (Gurtz et al. 1980). Accumulated sediments in the stream channel were greatest in low-gradient, sandy-gravel reaches > pebble riffles > cobble riffles and least in the steep gradient moss-covered bedrock substrates (Gurtz and Wallace 1984). Furthermore, BHB substrates in the sandy-gravel, pebble, and cobble habitats contained significantly higher proportions of sand than did those of the reference stream HWC (Gurtz and Wallace 1984). Thus, it is questionable that even riparian zone protection would have provided adequate protection to elevated sediment loading from road building in WS 7. Sedimentation associated with the exceptional April 1976 rainfall, which coincided with recent road construction, may have overwhelmed the capacity of the riparian zone to protect BHB at the stream crossings.

Under normal conditions adequate riparian zones may have resulted in different results that those we observed for benthic macroinvertebrates. First, because of greater shading of the stream, there would not have been a large increase in primary production and stream temperature increases (Swift 1983) would not have been as prominent. Thus, the observed shift from an allochthonous to an autochthonous energy base of the invertebrate assemblage would have been reduced. Second, a larger source of allochthonous detritus to the stream bed would have been maintained, and the energy base of the stream would not have shifted from allochthonous detritus to autochthonous primary production as was observed by Webster et al. (1983). Finally, although most sediment input to the stream was associated with road building, a riparian management zone would have reduced fine-sediment input. Even a modest riparian buffer width (3 m) in association with moderate-intensity harvesting (42% basal area removed) has been shown to mitigate changes in canopy cover, temperature, water chemistry, fine-sediment input, and litter input with little change in macroinvertebrate assemblage structure (Kreutzweiser et al. 2005). In summary, the establishment of a riparian management zone would undoubtedly have reduced some of the major changes observed in invertebrate assemblage (i.e., increases in grazers such as baetids and the influx of taxa from downstream areas) and lessened the impact of sediments on fauna of the lower-gradient depositional zones (Quinn et al. 2004).

Another factor that should be considered is that BHB is a high-gradient, erosional stream with a diverse array of substrates ranging from depositional zones to extremely steep, moss-covered, bedrock outcrops. Clearcutting of low-gradient streams, such as those of the southeastern Coastal Plain, may yield somewhat different results. For example, Kedzierski and Smock (2001) studied the response of invertebrates to logging in a low-gradient, soft-bottomed stream in Virginia and found several points of contrast from high-gradient streams that had been logged. First, there was increased macrophyte production in logged reaches of low-gradient streams compared to increased periphyton production in the rocky reaches of high-gradient streams. Second, the abundant macrophytes resulted in much higher productivity among collectors, filterers, and predators in the logged reaches, whereas scrapers were the least affected. Thirdly, macrophytes appear to be a key factor in influencing invertebrate response to logging in low-gradient streams (Kedzierski and Smock 2001). Thus, geomorphology and landscape must be considered as important variables when assessing the response of invertebrates to watershed disturbances such as clearcutting.

Invariably, in long-term studies one confronts several decisions in field sampling for benthic invertebrates with regard to sampling methods, taxonomic considerations, and differences in sampling frequency over several decades. As improved and more efficient sampling methods evolve, for example, use of benthic corer and a laboratory sample splitter for more efficient recovery of small organisms, we should always strive for the most effective methods. Unfortunately, this may result in values for abundance, biomass, and production that are not directly comparable with earlier studies; however, it will give a better picture of the current conditions. In the last 25 years, more effective and complete keys to benthic invertebrates, especially the aquatic insects, have been developed (e.g., Merritt et al. 2008). Long-term storage of representative specimens from earlier studies and adequate institutional museum space and resources for maintaining specimens are required to assess potential changes in taxonomy. Finally, the resources available to conduct a given project will vary among studies, and researchers must decide the most appropriate methods and sampling regimes to best coincide with both the short- and long-term objectives of a given study.

Acknowledgments

This work was supported by grants from the National Science Foundation, including from the Long-Term Ecological Research Program. Numerous people helped with various aspects of the work including Drs. Marty Gurtz, Jackson Webster, Fred Benfield, and Francie Smith Cuffney, as well as J. Clark Miller, Joe O'Hop, Mike Stone, and others.

Literature Cited

Anderson, N. H. 1992. Influence of disturbance on insect communities in Pacific Northwest streams. *Hydrobiologia* 248: 79–92.

Banks, J. L., J. Li, and A. T. Herlihy. 2007. Influence of clearcut logging, flow duration, and season on emergent aquatic insects in headwater streams of the Central Oregon Coast Range. *Journal of the North American Benthological Society* 26: 620–632.

Bender, E. A., T. J. Case, and M. E. Gilpin.1984. Perturbation experiments in community ecology: theory and practice. *Ecology* 65: 1–13.

Benke, A. C. 1993. Concepts and patterns of invertebrate production in running waters. *Internationale Vereinigung für Theoretische und Angewandte Limnologie, Verhandlungen* 25: 15–38.

Benke, A. C., and D. I. Jacobi. 1994. Production dynamics and resource utilization of snag-dwelling mayflies in a blackwater river. *Ecology* 75: 1219–1232.

Benke, A. C., T. C. Van Arsdall, Jr., D. M. Gillespie, and F. K. Parish. 1984. Invertebrate productivity in a subtropical blackwater river: the importance of habitat and life history. *Ecological Monographs* 54: 25–63.

Benke, A. C., and J. B. Wallace.1980. Trophic basis of production among net-spinning caddisflies in a southern Appalachian stream. *Ecology* 61: 108–118.

Cross, W. F., J. B. Wallace, A. D. Rosemond, and S. L. Eggert. 2006. Whole-system nutrient enrichment increases secondary production in a detritus-based ecosystem. *Ecology* 87: 1556–1565.

Cross, W. F., J. B. Wallace, and A. D. Rosemond. 2007. Nutrient enrichment reduces constraints on material flows in a detritus-based food web. *Ecology* 88: 2563–2575.

Edington, J. M., and A. H. Hildrew. 1973. Experimental observations relating to the distribution of net-spinning Trichoptera in streams. *Internationale Vereinigung für Theoretische und Angewandte Limnologie, Verhandlunge* 18: 1549–1558.

Ely, D. T. 2005. Long-term response of stream invertebrates to catchment logging. MS thesis. University of Georgia, Athens.

Ely, D. T., and J. B. Wallace. 2010. Long-term functional group recovery of lotic macroinvertebrates from logging disturbance. *Canadian Journal of Fisheries and Aquatic Sciences* 67: 1126–1134.

Feminella, J. W., and C. P. Hawkins. 1995. Interactions between stream herbivores and periphyton: a quantitative analysis of past experiments. *Journal of the North American Benthological Society* 14: 465–509.

Fortino, K., A. E. Hershey, and K. J. Goodman. 2004. Utility of biological monitoring for detection of timber harvest effects on streams and evaluation of Best Management Practices: a review. *Journal of the North American Benthological Society* 23: 634–646.

Fuchs, S. A., S. G. Hinch, and E. Mellina. 2003. Effects of streamside logging on stream macroinvertebrate communities and habitat in the sub-boreal forests of British Columbia, Canada. *Canadian Journal of Forest Research* 33: 1408–1415.

Fuller, R. L., J. R. Roelofs, and T. J. Fry. 1986. The importance of algae to stream invertebrates. *Journal of the North American Benthological Society* 5: 290–294.

Golladay, S. W., J. R. Webster, and E. F. Benfield. 1989. Changes in stream benthic organic matter following watershed disturbance. *Holarctic Ecology* 12: 96–105.

Grubaugh, J. W., J. B. Wallace, and E. S. Houston. 1997. Production of benthic macroinvertebrate communities along a southern Appalachian river continuum. *Freshwater Biology* 37: 581–596.

Gurtz, M. E. 1981. Ecology of stream invertebrates in a forested and commercially clear-cut watershed. PhD dissertation. University of Georgia, Athens.

Gurtz, M. E., and J. B. Wallace. 1984. Substrate-mediated response of stream invertebrates to disturbance. *Ecology* 65: 1556–1569.

Gurtz, M. E., and J. B. Wallace. 1986. Substrate-production relationships in net-spinning caddisflies (Trichoptera) in disturbed and undisturbed hardwood catchments. *Journal of the North American Benthological Society* 5: 230–336.

Gurtz, M. E., J. R. Webster, and J. B. Wallace. 1980. Seston dynamics in southern Appalachian streams: effects of clearcutting. *Canadian Journal of Fisheries and Aquatic Sciences* 37: 624–631.

Hains, J. J. 1981. The response of stream flora to watershed perturbation. MS thesis. Clemson University. Clemson, South Carolina.

Hall, R. O., G. E. Likens, and H. M. Malcolm. 2001. Trophic basis of invertebrate production in 2 streams at the Hubbard Brook Experimental Forest. *Journal of the North American Benthological Society* 20: 432–447.

Hawkins, C. P., M. L. Murphy, and N. H. Anderson. 1982. Effect of canopy, substrate composition, and gradient on the structure of macroinvertebrate communities in Cascade Range streams of Oregon. *Ecology* 63: 1840–1856.

Huryn, A. D. 1998. Ecosystem-level evidence for top-down and bottom-up control of production in a grassland stream system. *Oecologia* 115: 173–183.

Huryn, A. D., and J. B. Wallace. 1986. A method for obtaining *in situ* growth rates of larval Chironomidae (Diptera) and its application to secondary production estimates. *Limnology and Oceanography* 31: 216–222.

Haefner, J. D., and J. B. Wallace. 1981. Shifts in aquatic insect populations in a first-order southern Appalachian stream following a decade of old field succession. *Canadian Journal of Fisheries and Aquatic Sciences* 38: 353–359.

Jackson, C. R., D. P. Batzer, S. S. Cross, S. M. Haggerty, and C. A. Sturm. 2007. Headwater streams and timber harvest: Channel, macroinvertebrate, and amphibian response and recovery. *Forest Science* 53: 356–370.

Karr, J. R. 1994. Biological monitoring: challenges for the future. Pages 357–373 in *Biological Monitoring of Aquatic Systems*. S.L. Loeb and A. Spacie, editors. Lewis Publishers, Boca Raton, Florida.

Kedzierski, W. M., and L. A. Smock. 2001. Effects of logging on macroinvertebrate production in a sand-bottomed, low-gradient stream. *Freshwater Biology* 46: 821–833.

Kreutzweiser, D. P., S. S. Capell, and K. P. Good. 2005. Macroinvertebrate community responses to selection logging in riparian and upland areas of headwater catchments in a northern hardwood forest. *Journal of the North American Benthological Society* 24: 208–222.

Krueger, C. C., and T. F. Waters. 1983. Annual production of macroinvertebrates in three streams of different water quality. *Ecology* 64: 840–850.

Lenat, D. R. 1993. A biotic index for the southeastern United States: derivation and list of tolerance values, with criteria for assigning water-quality ratings. *Journal of the North American Benthological Society* 12: 279–290.

Martel, N., M. A. Rodriquez, and P. Berube. 2007. Multi-scale analysis of responses of stream macrobenthos to forestry activities and environmental context. *Freshwater Biology* 52: 85–97.

Meyer, J. L., and C. M. Tate.1983. The effects of watershed disturbance on dissolved organic carbon dynamics of a stream. *Ecology* 64: 33–44.

Merritt, R. W., K. W. Cummins, and M. B. Berg. 2008. An introduction to the aquatic insects of North America. 4th ed. Kendall/Hunt, Dubuque, Iowa.

Newbold, J. D., D. C. Erman, and K. B. Roby. 1980. Effects of logging on macroinvertebrates in streams with and without buffer strips. *Canadian Journal of Fisheries and Aquatic Sciences* 37: 1076–1085.

Niemi, G. J., N. Detenbeck, and J. Perry. 1993. Comparative analysis of variables to measure recovery rates in streams. *Environmental Toxicology and Chemistry* 12: 1541–1547.

Nislow, K. H., and W. H. Lowe. 2006. Influences of logging history and riparian forest characteristics on macroinvertebrates and brook trout (*Salvelinus fontinalis*) in headwater streams (New Hampshire, U.S.A.). *Freshwater Biology* 51: 388–397.

Noel, D. S., C. W. Martin, and C. A. Federer. 1986. Effects of forest clearcutting in New England USA on stream macroinvertebrates and periphyton. *Environmental Management* 10: 661–670.

Quinn, J. M., I. K. G. Boothroyd, and B. J. Smith. 2004. Riparian buffers mitigate effects of pine plantation logging on New Zealand streams 2. Invertebrate communities. *Forest Ecology and Management* 191: 129–146.

Reice, S. R., and M. Wohlenberg. 1993. Monitoring freshwater benthic macroinvertebrates and benthic processes: Measures for assessment of ecosystem health. Pages 287–305 in *Freshwater Biomonitoring and Benthic Macroinvertebrates*. V. Resh and D. Rosenberg, editors. Chapman and Hall, New York, New York.

Ross, H. H. 1963. Stream communities and terrestrial biomes. *Archiv für Hydrobiologie* 59: 235–242.

Stone, M. K., and J. B. Wallace. 1998. Long-term recovery of a mountain stream from clear-cut logging: the effects of forest succession on benthic invertebrate community structure. *Freshwater Biology* 39: 141–169.

Stout, B. M., E. F. Benfield, and J. R. Webster. 1993. Effects of forest disturbance on shredder production in southern Appalachian streams. *Freshwater Biology* 29: 59–69.

Swank, W. T. 1988. Stream chemistry response to disturbance. Pages 339–357 in *Forest Hydrology and Ecology at Coweeta*. W. T. Swank and D. A. Crossley, Jr., editors. Springer-Verlag, New York, New York.

Swank, W. T., J. E. Douglass, and G. B. Cunningham. 1982. Changes in water yield and storm hydrographs following commercial clearcutting on a southern Appalachian catchment. Pages 583–594 in *Hydrological Research Basins and Their Use in Water Resource Planning*. Swiss National Hydrological Service Special Publication, Berne, Switzerland.

Swank, W. T., J. M. Vose, and K. J. Elliot. 2001. Long-term hydrologic and water quality responses following commercial clearcutting of mixed hardwoods on a southern Appalachian catchment. *Forest Ecology and Management* 143: 163–178.

Swank, W. T., and J. B. Waide. 1988. Characterization of baseline precipitation and stream chemistry and nutrient budgets for control watersheds. Pages 57–76 in *Forest Hydrology and Ecology at Coweeta*. W. T. Swank and D. A. Crossley, Jr., editors. Springer-Verlag, New York, New York.

Swift, L.W. 1983. Duration of stream temperature increases following forest cutting in the southern Appalachian Mountains. Pages 273–275 in *Proceedings of the International Symposium on Hydrometeorology*. A. I. Johnson and R. A. Clark, editors. American Water Resources Association, Bethesda, Maryland.

Tank, J. L., and J. R. Webster. 1998. Interaction of substrate availability and nutrient distribution on wood biofilm development in streams. *Ecology* 79: 2168–2179.

Wallace, J. B. 1987. Aquatic invertebrate research at Coweeta. Pages 257–268 in *Forest Hydrology and Ecology at Coweeta*. W. T. Swank and D. A. Crossley, Jr., editors. Springer-Verlag, New York, New York.

Wallace, J. B., and J. R. Webster. 1996. The role of macroinvertebrates in stream ecosystem function. *Annual Review of Entomology* 41: 115–139

Wallace, J. B., S. L. Eggert, J. L., Meyer, and J. R. Webster. 1997. Multiple trophic levels of a forest stream linked to terrestrial litter inputs. *Science* 277: 102–104.

Wallace, J. B., S. L. Eggert, J. L. Meyer, and J. R. Webster. 1999. Effects of resource limitation on a detrital-based ecosystem. *Ecological Monographs* 69: 409–442.

Wallace, J. B., and M. E. Gurtz. 1986. Response of Baetis mayflies (Ephemeroptera) to catchment logging. *American Midland Naturalist* 115: 25–41.

Wallace, J. B., M. E. Gurtz, and F. Smith-Cuffney. 1988. Long-term comparisons of insect abundances in disturbed and undisturbed Appalachian headwater streams. *Internationale Vereinigung für Theoretische und Angewandte Limnologie, Verhandlungen* 23: 1224–1231.

Wallace, J. B., J. W. Grubaugh, and M. R. Whiles. 1996. Biotic indices and stream ecosystem processes: results from an experimental study. *Ecological Applications* 6: 140–151.

Webster, J. R., M. E. Gurtz, J. J. Hains, J. L. Meyer, W. T. Swank, J. B. Waide, and J. B. Wallace. 1983. Stability of stream ecosystems. Pages 355–395 in *Stream Ecology*. J. R. Barnes and G. W. Minshall, editors. Plenum Press, New York, New York.

Webster, J. R., and J. B. Waide. 1982. Effects of forest clearcutting on leaf breakdown in a Southern Appalachian stream. *Freshwater Biology* 12: 331–344.

Yount, J. D., and G. J. Niemi. 1990. Recovery of lotic communities and ecosystems from disturbance: a narrative review of case studies. *Environmental Management* 14: 547–569.

Zweig, L. D., and C. F. Rabeni. 2001. Biomonitoring for deposited sediment using benthic invertebrates: a test on 4 Missouri streams. *Journal of the North American Benthological Society* 20: 643–657.

12

Recovery of Central Appalachian Forested Watersheds
Comparison of Fernow and Coweeta

Mary Beth Adams*
James N. Kochendenfer

Introduction

The Fernow Experimental Forest (FEF) was established to conduct research in forest and watershed management in the central Appalachians. The 1868-ha FEF, located south of Parsons, West Virginia, is administered by the Northern Research Station of the USDA Forest Service and provides a valuable point of comparison with Coweeta Hydrologic Laboratory (CHL), located in the southern Appalachians. This chapter summarizes responses to clearcutting on four watersheds at FEF and compares the results to those from clearcutting on CHL Watershed 7 (WS 7).

The Elklick watershed (which later became the bulk of the FEF) was initially logged between 1903 and 1911 during the railroad-logging era (Trimble 1977). Wind is considered the dominant disturbance agent on the Fernow, but early snow, when leaves are still on some trees, has also been an important disturbance. Forest fires may have been an important disturbance agent prior to initial logging, but Bryant (1911), after examining the property in 1911, determined that there had been no fires on it for a long time. Most of the Elklick watershed was not farmed, and the forest was allowed to regenerate naturally following the cessation of logging activities. The current mature forest developed in the absence of deer, with very low levels of herbivory (DeGarmo and Gill 1958; Kochenderfer 1975). Chestnut blight was first noted in West Virginia as early as 1909 (Brooks 1911) and in places resulted in a 25% reduction in standing volume on the experimental forest in the 1930s (Weitzman,1949). More historical information was published by Kochenderfer (2006).

* Corresponding author: USDA Forest Service, P.O. Box 404, Parsons, WV 26287 USA .

Site Description

The ecological land type of the FEF is referred to as the Allegheny Mountains section of the Central Appalachian Broadleaf Forest, according to the Forest Service National Hierarchical Framework of Ecological Units (McNab and Avers 1994). The land-type association is designated as Allegheny Front Side Slopes, and vegetation is classified as mixed mesophytic. Characteristic species include northern red oak (*Quercus rubra*), yellow poplar (*Liriodendron tulipifera*), black cherry (*Prunus serotina*), sugar maple (*Acer saccharum*), sweet birch (*Betula lenta*), red maple (*A. rubrum*), and American beech (*Fagus grandifolia*). Leaf area index for mature forest on good to excellent sites is 4.5.

The topography is mountainous, with elevations ranging from 530 to 1115 m above sea level. Mean annual precipitation is about 1,480 mm, distributed evenly throughout the year. The growing season is May through October with an average total frost-free period of 145 days. Snow is common in winter, but a snowpack generally lasts no more than a few weeks; snow contributes approximately 14% of the precipitation to FEF (Adams et al. 1994). Mean annual temperature is 9.2°C but temperatures reach −20°C most winters. Large rainfall events are normally associated with hurricanes. About half of the largest storms on the Fernow have occurred during the dormant season (November 1–April 30; Adams et al. 1994), when evapotranspiration losses are low. The largest peak flow (0.72 m³/s) recorded on FEF4 occurred in November 1985, after a 15.24-cm rainfall.

Slopes ranging from 20% to 50% cover most of the area. The soils are predominantly Inceptisols from the Calvin and Dekalb soil series. The Calvin series consists of moderately deep, well-drained soils formed in material weathered from interbedded shale, siltstone, and sandstone. Dekalb soils are also acidic, deriving from acidic sandstones. Average soil depth is about 1 m, and the soil contains a considerable amount of stones and large gravels.

Predictions: Fernow in Comparison and Contrast to Coweeta

There are many similarities between the two research locations, and some striking differences. The forest of the FEF is mesic, mixed hardwood, similar to the cove-hardwood and mixed-oak hardwood forests at CHL, but with some significant differences in species composition. For example, black locust (*Robinia pseudoacacia*) is an important part of early successional forests at CHL. Black locust is common at FEF but not in such abundance as at CHL, nor is the very high early mortality of black locust observed at CHL (55%; Elliott et al. 1997) so evident at FEF. Rhododendron (*Rhododendron maximum*) is more abundant at CHL, particularly in the riparian zone, than at FEF. Black cherry is much more abundant at FEF than at CHL. We therefore hypothesize that there may be differences in transpiration rates and nutrient cycling due to these species' physiological characteristics. Soils are also generally deeper at CHL, suggesting some differences in hydrologic characteristics, particularly soil moisture and storage. Because CHL receives more rainfall (~ 2000 mm/yr compared to 1500 mm/yr at FEF), soils at CHL are subjected to more leaching and are more well-developed than those at FEF.

Atmospheric deposition of nutrients historically has been and continues to be greater at FEF: N deposition is approximately 15 kg N ha^{-1} yr^{-1} at FEF compared with 4.5 kg N ha^{-1} yr^{-1} at CHL. Deposition of Ca and K to FEF is about twice that deposited at CHL, while sulfate deposition is approximately 40% greater at FEF than at CHL (www.nadp.sws.uiuc.edu/). These differences in deposition could contribute to significant differences in nutrient cycling and plant growth between the two locations. While we note these differences, we predict that the response to clearcutting will be similar between the two research forests. We expect to see the greatest differences in response to be relative to the cycling of nutrients because of the differences between the two regions in nutrient inputs, soil weathering, and growth of the forest vegetation.

Watershed Treatments at the Fernow

We examined the responses of 4 of the 10 gaged watersheds on the FEF: watersheds 1, 3, 6, 7 (FEF1, FEF3, FEF6, and FEF7, respectively). Each of these watersheds was clearcut, although at different times (table 12.1). FEF4 serves as the reference watershed for these 4 Fernow watersheds (figure 12.1). We compared our results with Coweeta's WS 7, a 59-ha watershed with a southern aspect that was commercially clearcut in 1977 (see Swank and Webster, chapter 1, this volume).

On the Fernow watersheds, stream discharge has been monitored using 120° sharp-crested V-notch weirs equipped with FW-1 water level recorders. FEF1, FEF3, and FEF4 have been gauged since May 1951; and FEF6 and FEF7, since November 1957.

Basic streamflow data presented here were determined from flow summaries. The hydrologic year begins on May 1 when the soil usually is fully recharged with moisture. For water yield determinations, growing and dormant seasons are designated to extend from May 1 to October 31 and from November 1 to April 30, respectively. Flow data have been analyzed as described by Reinhart et al. (1963) at $P \leq 0.05$. Stream water grab samples have been collected from FEF1 through FEF4 on a weekly or biweekly basis since 1960; and since 1971 on FEF6 and FEF7. Details of other measurements and analyses were given by Adams et al. (1994).

Results of Experimental Treatments at FEF

Forest Regeneration

Recovery of vegetation on FEF3, FEF6, and FEF7 began with the 1970 growing season. Natural plant succession on FEF6 and FEF7 began at the grass and herbaceous stage (Kochenderfer and Wendel 1983) as a result of the herbicide treatment;

Figure 12.1   Stream gaging station at Fernow Experimental Forest 4 (FEF4), reference watershed, during high flow. (USDA Forest Service photo)

whereas on FEF3 vegetation development began at a more advanced successional stage, and consisted mainly of woody vegetation and *Rubus* spp. Much of the regrowth on FEF3 consisted of sprouts utilizing existing root systems, while regrowth on FEF6 and FEF7 originated mostly from seed, making regrowth and reoccupation of the site slower. Norway spruce (*Picea abies*) was planted on FEF6 in 1973, and further herbicide treatment of competing hardwoods was needed to ensure occupation of the site by the spruce trees.

Total aboveground biomass on FEF7 increased to approximately 33 T/ha within the first 10 years after the end of the herbicide treatment, with 77% of that biomass being produced in the last three years (Kochenderfer and Wendel 1983). By 1991, aboveground biomass was 80 T/ha for FEF7 and 97 T/ha for FEF3, compared to 312 T/ ha for a mature (~90 years old) stand (Adams et al. 1995). Thus, within 30 years, FEF3 had accumulated approximately 53% of the biomass of a 90-year-old stand, and FEF7, 40% of the biomass of the mature stand. Average annual leaf fall mass, measured since 1989, did not vary significantly between FEF3 and FEF7 (Adams 2008), although it was significantly less than that from FEF4 (74% that of FEF4).

In 1999, FEF3 supported a young hardwood stand dominated by black cherry, red maple, American beech, and black birch, while FEF7 supported a young stand dominated by black birch, sugar maple, red maple, black cherry, and yellow poplar. In 1999, black cherry accounted for more than half of the basal area in trees 2.5 cm

Table 12.1   Description of Fernow watersheds and the treatments applied.

Watershed	Treatment	Treatment Date	Basal Area cut %	Aspect	Area (ha)
1	Clearcut to 15 cm d.b.h., except culls	May 57–June 58	74	NE	30.11
	Fertilized with 500 kg/ha urea	May 71			
3	Intensive selection cut, including culls in trees > 12.7 cm d.b.h.,	Oct. 58–Feb. 59	13	S	34.39
	Repeat treatment	Sept. 63–Oct. 63	8		
	0.16 ha patch cuttings totaling 2.3 ha, cut down to 12.6 cm, 2–12 cm stems sprayed with herbicide	July 68–Aug. 68	6		
	Clearcut to 2.5 cm d.b.h., except for a partially cut 3.0-ha shade strip along the stream channel	July 69–May 70	91		
	Shade strip clearcut	Nov. 72	9		
	Ammonium sulfate fertilizer applied,	Dec. 89–present			
4	Reference	None		ESE	38.73
6	Lower 11 ha clearcut	Mar 64–Oct. 64	51	S	22.34
	Maintained barren w/ herbicides	May 65–Oct. 69			
	Upper 11 ha clearcut	Oct. 67–Feb. 68	49		
	Entire watershed maintained barren with herbicides	May 68–Oct. 69			
	Planted with Norway spruce	Mar. 73			
	Aerially spray with herbicides	Aug. 75, Aug. 80			
7	Upper 12 ha clearcut	Nov. 63–Mar. 64	49	E	24.23
	Maintained barren with herbicides	May 64–Oct. 69			
	Lower 12 ha clearcut	Oct. 66–Mar. 67	51		
	Entire watershed maintained barren with herbicides	May 67–Oct. 69			

and larger on FEF3. The dominant trees on good sites on FEF1 in 1995 were sugar maple, yellow poplar and basswood. The percentage of yellow poplar and sugar maple basal area on FEF1 increased 7% and 8%, respectively, in 1995 from the original inventory in 1958, while the percentage of hickory basal area decreased from 10% to 0% and northern red oak from 13% to 7% on good sites. This decrease in large-seeded species was also observed on FEF3 and on other areas across the Fernow (Schuler and Gillespie 2000). An increase in shade-intolerant tree species and a decrease in large-seeded and shade-tolerant species was also reported for Coweeta (Elliott et al. 1997). We cannot attribute these species changes solely to the clearcutting treatments, however. The species composition of a stand is a complex issue, reflecting factors such as past land-use history, disturbance history, deer browsing, seed predation, insects, and disease. For example, small canopy gaps in the overstory combined with recent high deer density and no control of shade tolerant species in the understory on the Fernow has heavily favored red maple, sugar maple, and an understory of American beech and striped maple at the expense of most other species (Kochenderfer 2006).

Water Yield and Peakflow

Figures 12.2 through 12.5 depict deviation of water yields from the predicted flows for FEF1, FEF3, FEF6, and FEF7, using prediction equations developed during the appropriate calibration period. Effects of the harvesting treatments on streamflow have previously been summarized for these and other watersheds by Kochenderfer et al. (1990) and Hornbeck et al. (1993). Annual water yield increased immediately after cutting in these watersheds. The initial flow increases were generally greater

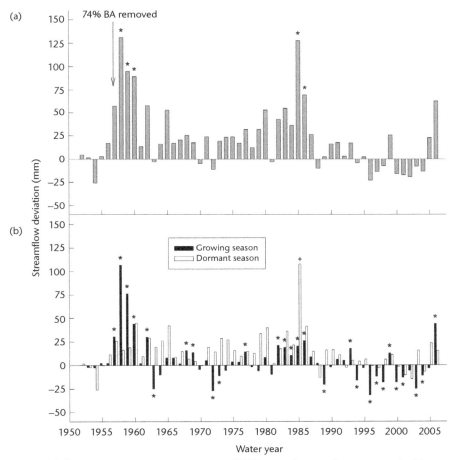

Figure 12.2   Fernow Experimental Forest 1 (FEF1) actual streamflow compared with predicted. (**A**) Annual water yield variation from predictions. Asterisk indicates statistically significant deviation from prediction ($P = 0.05$). (**B**) Growing season and dormant season streamflow variation from predicted values. Asterisk (*) indicates statistically significant deviations for growing season; plus indicates statistically significant deviation for dormant season flows ($P = 0.05$).

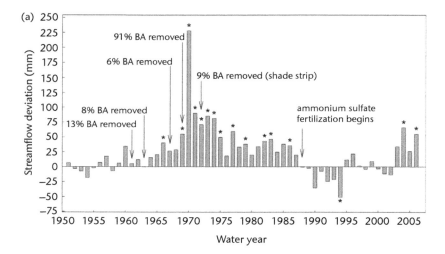

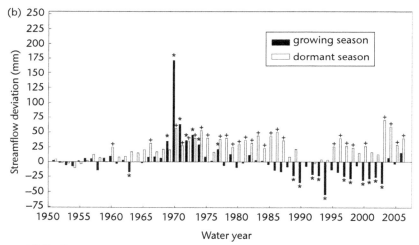

Figure 12.3   Fernow Experimental Forest 3 (FEF3) actual streamflow compared with predicted. (**A**) Annual water yield variation from predictions. Asterisk indicates statistically significant deviation from prediction ($P = 0.05$). (**B**) Growing season and dormant season streamflow variation from predicted values. Asterisks indicate statistically significant deviations for growing season; plus signs indicate statistically significant deviation for dormant season flows ($P = 0.05$).

during the growing season, suggesting that the increases in flow were largely due to reduced transpiration after cutting. Statistically significant increases in annual water yield from FEF3 over a longer time period reflected the additional removal of the streamside buffer. Use of herbicides on FEF6 and FEF7 to control regrowth also significantly prolonged increases in annual flow relative to FEF1. Both growing season and dormant season flows from FEF6 and FEF7 increased during the first 20–25 years after treatment (Kochenderfer et al. 1990), although these increases

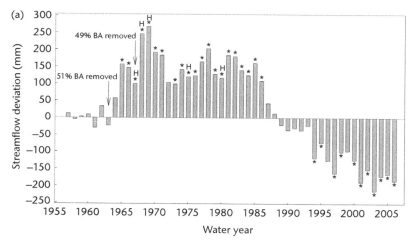

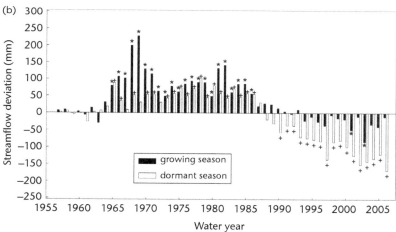

Figure 12.4   Fernow Experimental Forest 6 (FEF6) actual streamflow compared with predicted. *H* indicates herbicide treatments. (**A**) Annual water yield variation from predictions. Asterisk indicates statistically significant deviation from prediction ($P = 0.05$). (**B**) Growing season and dormant season streamflow variation from predicted values. Asterisks indicate statistically significant deviations for growing season; plus signs indicate statistically significant deviation for dormant season flows ($P = 0.05$).

were not always statistically significant. Note that this trend has changed and decreases in flow, relative to that predicted, have been observed on FEF1, FEF3, FEF6, and FEF7 since the 1990s (figures 12.2–5), although most differences were not statistically significant, except for FEF6.

Annual water yields for FEF1 returned to pretreatment levels within 4 years. Repeated disturbances to FEF3 (harvesting) and FEF6 and FEF7 (herbicides) appeared to extend statistically significant increases in annual yield to about 20–30 years post clearcutting. Note that statistically significant increases in annual yield were again

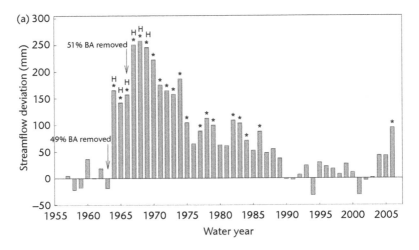

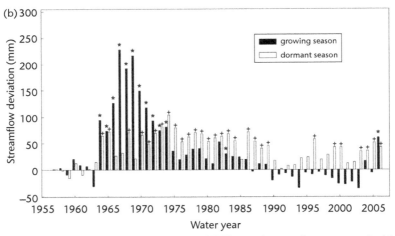

Figure 12.5   Fernow Experimental Forest 7 (FEF7) actual streamflow compared with predicted. *H* indicates herbicide treatments. (**A**) Annual water yield variation from predictions. Asterisk indicates statistically significant deviation from prediction ($P = 0.05$). (**B**) Growing season and dormant season streamflow variation from predicted values. Asterisks indicate statistically significant deviations for growing season; plus signs indicate statistically significant deviation for dormant season flows ($P = 0.05$).

detected for FEF1 in 1985 when dormant season flows were dramatically increased. A record storm in November 1985 (Kochenderfer et al. 2007) filled the weir pond on FEF1 with debris rendering streamflow measurement accuracies questionable during the storm. Hornbeck (1973) pointed to the problems of extrapolating extreme flow events. In addition, the steep unplanned road system used in 1957–58, closely associated with a high-gradient stream network, makes the FEF1 gaging station especially vulnerable to large debris flows during such unusual storms.

Significant increases in dormant season flow increases generally persisted longer for FEF3, FEF6, and FEF7 than for FEF1 and also generally longer than growing season increases for FEF3 and FEF7. The rapid decline in growing season water yield increases on FEF3 was attributed to luxuriant vegetative regrowth (Aubertin and Patric 1974). The lower than predicted growing season yields on FEF3 in the late 1980s through 2003, though not all statistically significant, could be due to the large increase in black cherry stems (from 5% to 50% of basal area) and to the fertilizer applications beginning in 1989, which coincided with the start of significant growing season declines. Black cherry consistently transpires at the highest rate per unit of leaf surface area found in hardwoods (Kochenderfer and Lee 1973). Also, a short-term growth response of black cherry to fertilization of FEF3 was observed (DeWalle et al. 2006). Therefore, some of the difference in growing season water yields between FEF3 and FEF7 during this time period could also be due to the greater importance of black cherry on FEF3 and to increased growth and transpiration due to the fertilization treatment. Hornbeck et al. (1993) advanced a similar hypothesis to explain effects of change in dominant species at Hubbard Brook—a significant increase in pin cherry (*Prunus pensylvanica*) and birch (*Betula allgeheniensis*) at the expense of beech and maple. Pin cherry has significantly lower leaf resistances, suggesting transpiration may be greater from a regrowing stand dominated by pin cherry and birch, with less water available for streamflow. Converting a hardwood-covered watershed at Coweeta (WS 6) to grass increased streamflow when the grass was not fertilized, but fertilization stimulated gross productivity and decreased streamflow to levels expected for the original hardwood forest (Swank et al. 1988).

Crown closure on FEF7 was delayed somewhat compared to FEF3, which may be attributed to the effect of the herbicide on regeneration sources. Because of the repeated herbicide treatments, stump sprouts were nearly eliminated on FEF7, and most regeneration originated from seeds (Kochenderfer and Wendel 1983). On FEF3, stump sprouts were the dominant regeneration source. Utilizing the existing rooting network on FEF3, the sprouts have had better access to soil moisture, resulting in greater transpiration at an earlier time than for FEF7.

Growing season water yield increases were longer lived for FEF6 than FEF7 because other vegetative regrowth (competing hardwoods) was controlled with aerial herbicide applications in 1975 and 1980 to release planted Norway spruce (Wendel and Kochenderfer 1984). Also, the planted spruce grew more slowly than the native hardwoods, and full site occupancy by the spruce required a longer time period. A survey in 1986 indicated that spruce crowns only covered about 24% of the ground area.

Reductions in annual water yield on FEF6 beginning in the 1980s can be attributed to the greater interception and transpiration, especially during the dormant season, by the planted conifer (Norway spruce) stand compared to the original hardwood stand (Helvey 1967; Delfs 1967). Annual streamflow reductions during the past 6 years on FEF6 have averaged 23%. Most of the significant decreases in FEF6 streamflow have occurred during the dormant season, when interception and transpiration by hardwood stands is low.

Streamflow is expected to continue to decline as the spruce stand matures. Delfs (1967) found that mean annual interception ranged from 21% in a 30-year-old Norway spruce stand in Germany to 36% in an 80-year-old stand. An estimate of mean annual hardwood interception (12.9%) was determined by applying Fernow precipitation data to dormant and growing season hardwood interception equations developed by Helvey and Patric (1965). Transpiration losses would also be expected to be much greater during the dormant season in the spruce stand. However, model simulations at Coweeta indicated that differences in annual interception and transpiration losses between white pine (*Pinus strobis*) and hardwood stands were about equal, despite greater dormant season transpiration by the white pine stand (Swank et al. 1988).

Dormant season peak flows on the harvested watersheds appeared little changed relative to the control watershed. This is attributed to the relatively small soil moisture deficits (higher soil moisture), because of low evapotranspiration during the dormant season. However, growing season peak flows were consistently higher on the clearcut watersheds where soil moisture deficits are reduced for a short period after cutting until vegetation regrows. This effect is more pronounced for the smaller storms, which provides support for the idea that differences in soil moisture are largely responsible for differences in growing season peak storm flows (Hornbeck et al. 1993).

The number of events considered to be storms increased with clearcutting (Bates 2000) due to increased soil moisture causing more response on clearcut watersheds. Because the relative increase is greater for small peaks, the number of events large enough to be considered storms is higher. Bates (2000) also reported that snowmelt peakflows appeared to occur and peak earlier on the FEF1 immediately after cutting relative to FEF4, probably due to greater net radiation on the snow cover, an effect also noted by Hornbeck (1970). Examination of hydrographs showed that, with the possible exception of snowmelt and excess runoff from logging roads when water was not controlled, there were no dramatic timing changes in the hydrographs after harvest, and subsurface flow was the main runoff production mechanism both before and after harvests.

Sediment Yields

Sediment yields prior to treatment and on the reference watershed ranged from 6 to 25 kg ha^{-1} yr^{-1} (Patric 1980; Kochenderfer et al. 1987). Clearcutting using an unplanned road system and no BMPs increased annual sediment yields to more than 3000 kg/ha on FEF1 during the logging operation (Kochenderfer and Hornbeck 1999) in 1957 and 1958, and to 97 kg/ha in 1970 for FEF3 where careful road management practices were followed. For both watersheds, within 5 years, annual sediment yield decreased rapidly to 44 and 28 kg/ha, respectively (Kochenderfer and Helvey 1984). Sediment yields are not available for FEF6 and FEF7, but deforestation of these watersheds did increase maximum turbidities observed during storm flows. However, nonstorm flows, constituting more than 90% of water yield, did not exceed 5 ppm of turbidity (Patric and Rinehart 1971). Most sediment was produced during storm flows (Kochenderfer et al. 1987). For all these studies, turbidity or suspended sediment returned to pretreatment or reference levels within a few years (Kochenderfer and Helvey 1984). Overland flow was seldom observed, only occurring on or directly below steeper roads (Patric 1973). Most of the sediment produced was delivered from roads, more rarely log landings, and the stream channels (Kochenderfer and Aubertin 1975).

Stream Temperature.

Clearcutting FEF1 raised stream temperature 4.5°C during the growing season and decreased temperature 2°C during the dormant season (Reinhart et al. 1963), and temperatures returned to pretreatment levels within 3 years. Eschner and Larmoyeux (1963) reported that clearcutting increased the maximum stream temperatures in summer and decreased the minimums in winter. There was a slight increase in growing season maximum temperatures for diameter-limit harvesting but no obvious effect of selection harvesting on stream temperatures. Clearcutting FEF3 in 1969 had no effect on temperature when a 50-foot-wide buffer strip was left along the channel. Removal of that buffer strip increased stream temperature about 4°C during the summer the shade strip was cut (Patric 1980). Channel shading was sufficient after 5 years of regrowth to return temperatures to preclearcutting levels (Patric 1980).

Stream Water Chemistry

Because of the relatively high levels of nitrogen deposition to the Fernow watersheds (Adams et al. 1993), the high rates of nitrification in the soil (Gilliam et al. 1996), and increasing levels of nitrogen emissions nationally, stream water nitrate concentrations are of particular interest. Stream water nitrate concentrations for the 4 watersheds are shown in figure 12.5. For all of these watersheds, only post-disturbance nutrient concentration data exist, with the exception of limited pre-treatment data on FEF1. Therefore statistical analyses of pre- and posttreatment differences are not feasible. However, several trends are particularly striking from even a quick glance at figure 12.6. In particular, the relatively high initial nitrate concentrations for FEF1, FEF6, and FEF7 are notable. The nitrate values for FEF1 reflect a fertilization with 500 kg/ha of urea in 1971. Prefertilization monthly maximum stream concentrations of nitrate-N were less than 2 mg/L, which increased to 16 mg/L immediately after fertilization (Kochenderfer and Aubertin 1975). Patric and Smith (1978) measured streamwater nitrogen and reported an annual loss of 25 kg/ha immediately after fertilization. The relatively high nitrate-N values recorded for FEF6 and FEF7 occurred 2 years after cessation of herbiciding. Clearcutting alone (FEF3) did not result in such large changes in stream nitrate-N or in any other chemical constituents (Aubertin 1971). FEF3 nitrate-N losses were less than 3 kg ha⁻¹ yr⁻¹ during the first 4 years after clearcutting, primarily because of rapid vegetative regrowth, retention of a lightly cut streamside zone, and good road management (Patric 1980). These study results demonstrate the importance

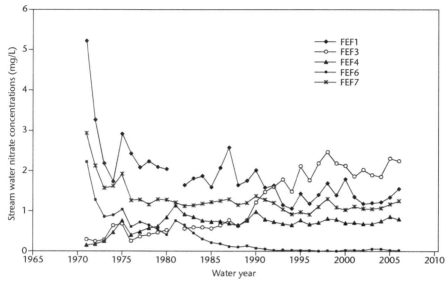

Figure 12.6 Stream water nitrate concentrations from five watersheds on the Fernow Experimental Forest, West Virginia. See table 12.1 for treatment descriptions.

of vegetation in maintaining water quality through nutrient uptake and control of microclimate.

Stream nitrate concentrations decreased quickly for FEF6 and FEF7 over the next 5 years to nearly the same level as FEF3 and FEF4. After 1983, nitrate concentrations from FEF6 decreased to near zero, while those of the other watersheds remained relatively constant, although FEF1 nitrate concentrations were consistently greater and more variable from year to year than those of the other watersheds. The extremely low nitrate concentrations recently observed on FEF6 may be due to increased interception along with preferential uptake of ammonium by the spruce trees and sequestration of nitrogen by an aggrading forest floor. Research is underway to elucidate the mechanisms. Nitrate concentrations increased in FEF3 as a result of fertilization with ammonium sulfate beginning in 1989 (Adams et al. 2006). In recent years, nitrate concentrations in FEF3 are approaching those observed on FEF6 and FEF7 immediately after deforestation.

The pattern for streamflow calcium concentrations is similar to that of nitrate concentrations for most streams (figure 12.7). The leaching of base cations is linked with the strong acid anions, particularly nitrate (Adams et al. 2006). Stream water magnesium concentrations are much lower than calcium concentrations, but the relative ranking of the watersheds by concentrations are the same as for calcium. Stream water sulfate concentrations showed no consistent pattern related to the cutting or herbicide treatments. Stream water pH has remained unchanged except on FEF3, where, as a result of fertilization, pH has decreased from 6.0 to 5.5.

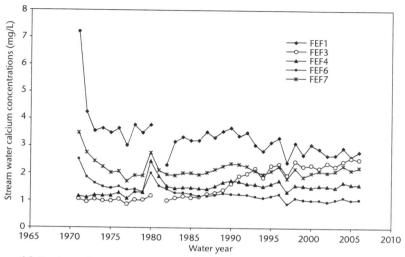

Figure 12.7   Annual stream water calcium concentrations from five watersheds on the Fernow Experimental Forest, West Virginia. See table 12.1 for treatment descriptions.

Comparisons with CHL Watershed 7

A comparison of results from Fernow and Coweeta clearcut watersheds reveals many similarities and a few differences. Hardwood forests regenerated quickly at both locations but slightly more quickly at CHL. By age 17 years at CHL, the stand had recovered most of its original basal area (Elliott et al. 1997), whereas this had occurred by 21 years at the FEF. Volume on FEF3 at 34 years was 65% of the precutting volume. Regeneration trajectories differed somewhat between the sites but were qualitatively similar. In the near term, clearcutting was found to favor shade-intolerant species, concomitant with a decrease in oaks, hickories, and shade-tolerant species at both locations. Many of the same tree species are common to both locations, but there are differences in relative proportions, mainly in the abundance of black locust, mountain laurel (*Kalmia latifolia* L.) and rhododendron (more common on Coweeta) and black cherry (more common at Fernow).

This comparison supports the conclusions of early research at Coweeta and Fernow, as well as other small watershed studies in the eastern United States, that increases in annual water yield could be expected from clearcutting, although the actual amount varied. Patric and Reinhart (1971) reported first-year water yield increases of 30 cm, compared with 41 cm in North Carolina and 33 cm at Hubbard Brook. Although the amounts may vary, the same pattern generally holds true over the long term: a rapid increase in annual water yields after clearcutting hardwood forested watersheds, followed by a quick return to pretreatment levels as revegetation occurs. However, there is a notable difference in water yield results between some of the Fernow watersheds and CHL WS 7. Significant increases in annual water yield seem to be of longer duration at three of the watersheds at Fernow (FEF3, 15 yr; FEF6, 20 yr; FEF7, 20 yr) than reported for CHL WS 7 or for FEF1(~ 4 yr). Each of the Fernow watersheds with longer recovery times received repeated vegetation removal treatments (whether by cutting or herbicides) as opposed to the single clearcuts on CHL WS 7 and FEF1. Hornbeck et al. (1993) identified intermediate cuttings and repeated herbicide use as contributors to prolonged streamflow increases.

For CHL WS 7, the largest flow increases occurred during the growing season; this was initially true for all the Fernow watersheds, providing evidence of the importance of transpiration in these forests' water balance. A few years after harvesting, however, on FEF3, FEF6, and FEF7, dormant season flows were significantly increased and sustained for a longer period of time relative to FEF1 and CHL WS 7. The reasons for this difference are not fully known but may be partially attributed to effects of the repeated treatments on evapotranspiration and consequent effects on soil moisture storage. It also could be due to differences in climate during the calibration period and during the intervening years. For example, cooler temperatures during the calibration period might suggest a larger proportion of dormant season precipitation came in the form of rain rather than snow during the treatment period. Consequently, evaporative losses and soil water content would be much smaller than predicted. We will continue to investigate these discrepancies.

Lessons learned from research on sediment yield and erosion are consistent across the two sites. Generally, overland flow does not occur in forested watersheds except on exposed roads where water was not properly controlled. Harvesting alone does not usually result in increased erosion or sediment inputs to streams. Carefully planned and prepared road systems, and use of Best Management Practices can minimize erosion and sediment inputs to streams.

The differences in stream chemistry between Fernow and Coweeta are probably due to the greater atmospheric inputs over a longer time period, particularly of nitrogen, at Fernow. At Fernow, streamwater concentrations of nitrate and calcium are much higher than CHL, but we did not see such a large relative increase after only clearcutting (FEF3)—on CHL WS 7 nitrate increased threefold or more as a result of clearcutting. However, this may be partly due to the very low background levels on CHL WS 7 (near detection limits), which connotes a more sensitive system. Converting hardwood watersheds to white pine at CHL also resulted in elevated nitrate concentrations up to 25 years later (Swank et al. 1988), whereas converting an FEF hardwood watershed to spruce resulted in significantly lower nitrate concentrations after an equal period of time.

It has been suggested that FEF4 is the best example of a "naturally" nitrogen-saturated watershed (Peterjohn et al. 1997), and this watershed has been used as an example of Stage 2 of nitrogen saturation (Stoddard 1994), whereas CHL WS 7 is considered to be in the latter phases of Stage 1 of nitrogen saturation (Swank and Vose 1997). At the Fernow, the largest increases in streamwater nitrate occurred when herbicide was used to prevent revegetation. This is similar to results from Hubbard Brook (see Bormann et al. 1968, chapter 17, this volume). Such results are not surprising, as inhibiting revegetation significantly decreases nutrient uptake and simultaneously increases water content and potentially water movement through the soil. Also by preventing revegetation, soil temperatures are elevated, increasing rates of decomposition and nutrient cycling. However, unless revegetation is prevented or delayed, as with herbicides, these cutting-induced peaks in nutrient concentrations generally are relatively short-lived. Results from both locations (FEF and CHL) suggest that elevated ecosystem nitrate availability, whether through atmospheric deposition or biological nitrogen fixation, can increase leaching of nitrate from forested watersheds. However, we can also conclude that in general, clearcutting did not affect nutrient concentrations to the extent of adversely affecting water quality for downstream users.

Conclusions

Comparisons between Fernow and Coweeta clear-cut watersheds reveal a number of consistencies:

- Regeneration/revegetation of harvested watersheds occurred rapidly.
- Clearcutting generally caused short-term increases in annual streamflow but generally had no effect on large peakflows.
- Repeated cuttings or devegetation using herbicides prolonged flow increases.
- Changes in species composition or species conversions can alter streamflow, but the duration of the effects may vary with successional trajectories.
- Nutrient losses increased after clearcutting, but the effects are variable depending on the intensity of the disturbance and the length of time revegetation requires.
- Sediment losses from clearcutting can be minimized through careful planning and use of Best Management Practices.

The differences observed between the two sites were relatively small and mostly dealt with rates of revegetation and nutrient cycling. These are probably due to differences in climate, atmospheric deposition, and soil depth.

The comparison of such research studies provides important opportunities to identify commonalities and differences and improve our understanding of forest ecosystem processes over a long timescale. The two research sites, Fernow and Coweeta, complement each other and provide valuable opportunities for broadening the conclusions of small watershed research through these comparisons. Such comparisons also speak to the importance of continuing such long-term watershed studies. As new questions and problems arise, we can use such long-term research in new contexts to further our understanding of ecosystems to help us address these new challenges. Finally, because trees are such long-lived organisms and forest ecosystems are dynamic in time and space, it is important to continue research throughout the life cycles our forests experience.

Acknowledgments

The assistance of Frederica Wood with data preparation and graphics assistance is gratefully acknowledged.

Literature Cited

Adams, M. B. 2008. Long-term leaf fall mass from 3 watersheds on the Fernow Experimental Forest, West Virginia. Pages 179–186 in *Proceedings, 16th Central Hardwood Forest Conference*. D. F. Jacobs and C. H. Michler, editors. Gen. Tech. Rep. NRS-P-24. USDA Forest Service, Northern Research Station, Newtown Square, Pennsylvania.

Adams, M. B., D. R. DeWalle, and J. L. Hom, editors. 2006. *The Fernow Watershed Acidification Study*. Springer, Dordrecht, The Netherlands.

Adams, M. B., P. J. Edwards, F. Wood, and J. N. Kochenderfer. 1993. Artificial watershed acidification on the Fernow Experimental Forest, USA.. *Journal of Hydrology* 150: 505–519.

Adams, M. B., J. N. Kochenderfer, F. Wood, T. R. Angradi, and P. J. Edwards. 1994. Forty years of hydrometeorological data from the Fernow Experimental Forest, West Virginia. Gen. Tech. Rep. NE-184. USDA Forest Service, Northeastern Forest Experiment Station, Radnor, Pennsylvania.

Adams, M. B., J. N. Kochenderfer, T. R. Angradi, and P. J. Edwards. 1995. Nutrient budgets of two watersheds on the Fernow Experimental Forest. Pages 119–130 in 10th Central Hardwood Forest Conference. K. W. Gottschalk and S. L. C. Fosbroke, editors. Gen. Tech. Rep. NE-107. USDA Forest Service, Northeastern Forest Experiment Station, Radnor, Pennsylvania.

Aubertin, G. M. 1971. Small watershed clearcutting is compatible with sound management of forest resources. 1971 Agronomy Abstracts. American Society of Agronomy, Madison, Wisconsin.

Aubertin, G. M., and J. H. Patric. 1974. Comments on water quality after clearcutting a small watershed in West Virginia. *Journal of Environmental Quality* 3: 243–249.

Bates, N. S. 2000. *Hydrologic effects of forest harvesting on headwater watersheds in West Virginia*. BS thesis. Princeton University, Princeton, New Jersey.

Bormann, F. J., G. E. Likens, D. W. Fisher, and R. S. Pierce. 1968. Nutrient loss accelerated by clearcutting of a forest ecosystem. *Science* 159: 882–884.

Brooks, A. B. 1911. Forestry and wood industries. *West Virginia Geologic and Economic Survey*, vol. 5. Acme Publishing Co., Morgantown, West Virginia.

Bryant, R.C. 1911. *Notes from report on preliminary examination of land offered by Thomas J. Arnold*. Fernow Experimental Forest files, Timber and Watershed Laboratory, Parsons, West Virginia.

DeGarmo, W. R., and J. Gill. 1958. *West Virginia white-tails*. Bulletin No. 4. WV Conservation Commission, Charleston, West Virginia.

Delfs, J. 1967. Interception and stemflow in stands of Norway spruce and beech in West Germany. Pages 179–185 in *International Symposium on Forest Hydrology*. W. E. Sopper and H. W. Lull, editors. Pergammon Press, London.

DeWalle, D. R., J. N. Kochenderfer, M. B. Adams, G. W. Miller, F. S. Gilliam, F. Wood, S. S. Odenwald-Clements, and W. E. Sharpe. 2006. Vegetation and Acidification. Pages 137–188 in *The Fernow Watershed Acidification Study*. M. B. Adams, D. R. DeWalle, and J. L. Hom, editors. Springer, Dordrecht, The Netherlands.

Elliott, K. J., L. R. Boring, W. T. Swank, and B. R. Haines. 1997. Successional changes in plant species diversity and composition after clearcutting a Southern Appalachian watershed. *Forest Ecology and Management* 92: 67–85.

Eschner, A. R., and J. Larmoyeux. 1963. Logging and trout: four experimental forest practices and their effect on water quality. *Progressive Fish-Culturist* 25: 59–67.

Gilliam, F. S., M. B. Adams, and B. M. Yurish. 1996. Ecosystem nutrient responses to chronic nitrogen inputs at Fernow Experimental Forest, West Virginia. *Canadian Journal of Forest Research* 16: 196–205.

Helvey, J. D. 1967. Interception by eastern white pine. *Water Resources Research* 3: 723–729.

Helvey, J. D., and J. H. Patrick. 1965. Canopy and litter interception of rainfall by hardwoods of eastern United States. *Water Resources Research* 1: 193–206.

Hornbeck, J. W. 1970. The radiant energy budget of clearcut and forested sites in West Virginia. *Forest Science* 16: 139–145.

Hornbeck, J. W. 1973. The problem of extreme events in paired-watershed studies. Res. Note NE-175. USDA Forest Service, Northeastern Research Station, Upper Darby, Pennsylvania.

Hornbeck, J. W., M. B. Adams, E. S. Corbett, E. S. Verry, and J. A. Lynch. 1993. Long-term impacts of forest treatments on water yield: a summary for northeastern USA. *Journal of Hydrology* 150: 323–344.

Kochenderfer, J. N. 2006. Fernow and the Appalachian hardwood region. Pages 17–39 in *The Fernow Watershed Acidification Study*. M. B. Adams, D. R. DeWalle, and J. L. Hom, editors. Springer, Dordrecht, The Netherlands.

Kochenderfer, J. N. 1975. Personal interviews with William Pennington, Parsons, West Virginia. Unpublished notes on file, Timber and Watershed Laboratory, Parsons, West Virginia.

Kochenderfer, J. N., M. B. Adams, G. W. Miller, and J. D. Helvey. 2007. Factors affecting large peakflows on Appalachian watersheds: Lessons from the Fernow Experimental Forest. Res. Pap NRS-3. USDA Forest Service, Northern Research Station, Newtown Square, Pennsylvania.

Kochenderfer, J. N., and G. M. Aubertin. 1975. Effects of management practices on water quality and quantity: Fernow Experimental Forest, West Virginia. Pages 14–24 in West Virginia Municipal Watershed Management symposium Proceedings. Gen. Tech. Rep. NE-13. USDA Forest Service, Northeastern Forest Experiment Station, Upper Darby, Pennsylvania.

Kochenderfer, J. N., P. J. Edwards, and J. D. Helvey. 1990. Land management and water yield in the Appalachians. Pages 523–532 in *Proceedings, IR Conference Watershed Management*, ID DIV/ASCE, Watershed Planning and Analysis in Action Symposium. R. E. Riggins, E. B. Jones, R. Singh, and P.A. Rechard, editors. American Society of Civil Engineers, New York, New York.

Kochenderfer, J. N., and J. D. Helvey. 1984. Soil losses from a "minimum-standard" truck road constructed in the Appalachians. Pages 215–225 in Proceedings, Mountain Logging Symposium. P. A. Peters and J. Luchok, editors. West Virginia University, Morgantown.

Kochenderfer, J. N., J. D. Helvey, and G. W. Wendel. 1987. Sediment yield as a function of land use in central Appalachian forests, Pages 497–502 in Proceedings, 6th Central Hardwood Forest Conference. R. L. Hay, F. W. Woods, and H. DeSelm, editors. University of Tennessee, Knoxville.

Kochenderfer, J. N., and J. W. Hornbeck. 1999. Contrasting timber harvesting operations illustrate the value of BMPs. Pages 128–135 in Proceedings, 12th Central Hardwood Forest Conference. J. W. Stringer and D. L. Loftis, editors. USDA Forest Service, Southern Research Station, Asheville, North Carolina.

Kochenderfer, J. N., and R. Lee. 1973. Indexes to transpiration by forest trees. *Oecologica Plantarum* 8: 175–184.

Kochenderfer, J. N., and G. W. Wendel. 1983. Plant succession and hydrologic recovery on a deforested and herbicided watershed. *Forest Science*. 29: 545–558.

McNab, W. H., and P. E. Avers. 1994. Ecological subregions of the United States: section descriptions. Administrative Publication WO-WSA5. USDA Forest Service, Washington, DC.

Patric, J. H. 1973. Deforestation effects on soil moisture, streamflow and water balance in the central Appalachians. Res. Pap. NE-259. USDA Forest Service, Northeastern Forest Experiment Station. Upper Darby, Pennsylvania.

Patric, J. H. 1980. Physical effects of timber harvesting on forest soil and water. Pages 1–3 in Proceedings of the Logging Residue Conference. M. L. Brooks, C. S. Hall, and J. Luchok, editors. West Virginia University, Morgantown.

Patric, J. H., and K. G. Reinhart. 1971. Hydrologic effects of deforesting two mountain watersheds in West Virginia. *Water Resources Research* 7: 1182–1188.

Patric, J. H., and D. W. Smith. 1978. Some effects of urea fertilization on a forested watershed in West Virginia. Pages 1–20 in Proceedings, 2nd Central Hardwood Forest Conference. Purdue University, West Lafayette, Indiana.

Peterjohn, W. T., M. B. Adams, and F. S. Gilliam. 1997. Symptoms of nitrogen saturation in two central Appalachian hardwood forest ecosystems. *Biogeochemistry* 35: 507–522.

Reinhart, K. G., A. Eschner, and G. R. Trimble, Jr. 1963. Effect on streamflow of four forest practices in the mountains of West Virginia. Res. Pap. NE-1. USDA Forest Service, Northeastern Forest Experiment Station. Upper Darby, Pennsylvania.

Schuler, T. M., and A. R. Gillespie. 2000. Temporal patterns of woody species diversity in a central Appalachian forest from 1856 to 1997. *Journal of the Torrey Botanical Society* 127: 149–161.

Stoddard, J. L. 1994. Long-term changes in watershed retention of nitrogen: Its causes and aquatic consequences. Pages 223–284 in Environmental chemistry of lakes and reservoirs. L. A. Baker, editor. American Chemical Society, Washington, DC.

Swank, W.T., L. W. Swift, Jr., and J. E. Douglass. 1988. Streamflow changes associated with forest cutting, species conversions, and natural disturbances. Pages 298–312 in *Forest Hydrology and Ecology at Coweeta*. W. T. Swank, and D. A. Crossely, Jr., editors. Springer-Verlag. New York, New York.

Swank, W. T., and J. M Vose. 1997. Long-term nitrogen dynamics of Coweeta forested watersheds in the southeastern United States of America. *Global Biogeochemical Cycles* 11: 657–671.

Trimble, G. R., Jr. 1977. A history of the Fernow Experimental Forest and the Parsons Timber and Watershed Laboratory. Gen. Tech. Rep. NE-28. USDA Forest Service, Northeastern Forest Experiment Station. Upper Darby, Pennsylvania.

Weitzman, S. 1949. The Fernow Experimental Forest. Miscellaneous Publication. USDA Forest Service, Northeastern Forest Experiment Station. Upper Darby, Pennsylvania.

Wendel, G. W., and J. N. Kochenderfer. 1984. Aerial release of Norway spruce with Roundup in the central Appalachians. *Northern Journal of Applied Forestry* 1: 29–32.

13

Comparisons with Results from the Hubbard Brook Experimental Forest in the Northern Appalachians

James W. Hornbeck*
Amey S. Bailey
Christopher Eagar
John L. Campbell

Introduction

The Hubbard Brook Experimental Forest (HBEF) is located in central New Hampshire, in the heart of the White Mountains, toward the northern end of the Appalachian chain. HBEF was established in 1955, two decades after Coweeta Hydrologic Laboratory (CHL), but research objectives at both sites have long been similar, that is, to understand hydrologic and nutrient cycling processes for forest ecosystems and to determine responses to natural and human disturbances.

This chapter summarizes the responses to intensive cuttings on three watersheds at HBEF and compares the results with those from the clearcutting on Watershed 7 (WS 7) at CHL. HBEF and CHL have some major differences in site characteristics that would be expected to cause variability between sites in hydrologic and nutrient cycles and responses to disturbances. Northern hardwood types with patches of spruce and fir at higher elevations dominate the forests at HBEF. The combination of diffuse porous species, including American beech, sugar maple, and yellow birch, have stomatal resistances, water use characteristics, and regeneration strategies that are different from those of the oak and oak-pine types found at CHL. The climate is cooler and drier at HBEF (annual precipitation averages 130 cm), and, in contrast to CHL, snow has an important role in winter and spring hydrology. Approximately one-third of annual precipitation at HBEF occurs as snow that accumulates as a snowpack. Runoff from the melting snowpack occurs primarily in April and accounts for 25% (20 cm) of the average annual streamflow (80 cm). The

* Corresponding author: USDA Forest Service, Northern Research Station, 271 Mast Road, Durham, NH 03824 USA

growing season is shorter at HBEF (May 15 to September 30); and complete soil moisture recharge occurs near the end of every autumn.

Metasedimentary and igneous rocks that are relatively impermeable to water from overlying soils underlie the watersheds at HBEF. Retreating glaciers approximately 14,000 years ago left deposits of unconsolidated till that vary widely in composition and depth. Soils are derived solely from this till and are predominantly Typic Haplorthods with sandy loam textures and high infiltration capacities. Soil depth is variable but seldom averages over 1 m. Thus soil rooting depth and water holding capacity are considerably less than at CHL. The glacial till and underlying geology weather slowly, and soils are relatively infertile, giving rise to streams that are clear and dilute (Hornbeck et al. 1997a).

Three treatments on gauged watersheds at HBEF can be contrasted with the clearcut logging experiment on CHL WS 7: (1) a clearfelling on HBEF Watershed 2 in winter 1965, with no roads or product removals, and application of herbicides for three successive summers after the felling operation; (2) a shelterwood harvest on HBEF Watershed 4 spanning 1970 to 1974, during which one-third of the watershed was harvested every other year by progressive strip cutting; and (3) a whole-tree clearcutting on HBEF Watershed 5 from October 1983 through May 1984 (figure 13.1). The methods used and detailed descriptions of these studies were published by Hornbeck et al. (1997b) and Martin et al. (2000). Soil disturbances ranged from little after the clearfelling and herbicide treatment to substantial after the whole-tree clearcutting. Canopy removal ranged from immediate during the clearfelling to gradual on the shelterwood. Regeneration was natural on all watersheds but ranged from uncontrolled after the whole-tree clearcutting and shelterwood harvests to controlled for 3 years after the clearfelling. Results are available for 41 years after the clearfelling, 36 years after the shelterwood, and 23 years after the whole-tree clearcutting. While all three watershed treatments at HBEF involved intensive cutting, the whole-tree clearcutting (Watershed 5) most closely resembles the clearcutting performed on WS 7 at CHL.

Results of Experimental Treatments at HBEF

The three watershed treatments at HBEF (figure 13.1) are hereafter referred to as CF (clearfelling and herbicide applications on Watershed 2), SC (shelterwood or progressive strip cutting on Watershed 4), and WT (whole-tree clearcutting on Watershed 5).

Forest Regeneration

Northern hardwoods regenerate by four major approaches: new seed, buried seed, stump sprouts, and root suckers (Hornbeck and Leak 1992). In addition some species also depend upon advanced regeneration already present in the understory at the time of disturbance. This variety of regeneration strategies provides a means of complete and reasonably rapid, natural revegetation of almost any type or size of disturbance. To demonstrate, table 13.1 compares mature versus regenerating

Figure 13.1   The Hubbard Brook experimental watersheds. The clearfelled watershed (CF) is right of center in this photo. The progressive strip-cut watershed (SC) is to the left of center and is shown with one-third of the strips harvested. The whole-tree harvest watershed (WT) is located to the left of the SC watershed and had not yet been harvested when this photo was taken. (USDA Forest Service photo)

Table 13.1   Basal area (m²/ha) by species for control and regenerating watersheds.

Species	Control mature[a]	CF, year 25[b]	SC, year 27[c]	WT, year 15[d]
Sugar maple (*Acer saccharum*)	9.1	1.0	2.1	0.4
Red maple (*Acer rubrum*)	0.2	0.5	0.3	0.1
Striped maple (*Acer pensylvanicum*)	0.1	0.4	1.1	0.6
American beech (*Fagus grandifolia*)	10.0	0.3	1.4	2.1
Yellow birch (*Betula alleghaniensis*)	5.3	2.4	7.6	3.1
Paper birch (*Betula papyrifera*)	1.7	4.0	2.7	2.3
White ash (*Fraxinus americana*)	0.3	0.5	1.1	0.1
Red spruce (*Picea rubens*)	0.7	< 0.05	0.1	0.1
Balsam fir (*Abies balsamea*)	0.7	0.7	0.2	0.7
Pin cherry (*Prunus pensylvanica*)	0.0	6.3	7.1	4.9
Trembling aspen (*Populus tremuloides*)	0.0	0.7	0.1	0.2
Others	0.2	0.3	0.5	0.2
Total	28.3	17.1	24.3	14.8

[a] 100% measurement in 1997 of all trees ≥ 10 cm DBH (data provided by T. G. Siccama, Yale University, New Haven, Connecticut)
[b] Measurements in 1991 of trees ≥ 2.5 cm DBH on 69 plots.
[c] Measurements in 1997 of trees ≥ 2.5 cm DBH on 57 plots
[d] Measurements in 1999 of trees ≥ 2.5 cm DBH on 101 plots (data provided by T.G. Siccama)

forests at HBEF. Despite the 3 years of herbicide applications, basal area of the CF watershed by year 25 after cutting had reached 17.1 m²/ha or 60% of the 28.3 m²/ha occurring on the mature forest of the control watershed. Regeneration grew even more rapidly on the SC, reaching an average basal area of 24.3 m²/ha at 27 years after initiation of cutting.

The forest treatments have caused a change in the species composition when compared with the control watershed. Eighty-six percent of the basal area on the control watershed consists of the primary northern hardwood species sugar maple (32%), American beech (35%), and yellow birch (19%) (table 13.1). In contrast, these same three species combine for only 22% of the basal area at 25 years after treatment on the CF watershed. Pin cherry (37%) and paper birch (23%), common pioneer species in northern hardwood forests, assumed early dominancy in the new forest. The same pattern occurred after SC and WT (table 13.1). These changes in species composition have important implications for water yield as discussed below.

Water Yield and Peakflow Rates

All three experimental treatments at HBEF were severe in that each nearly eliminated basal area and transpiration. However, responses in annual water yields varied markedly among treatments (table 13.2). CF produced the most dramatic response, causing annual water yields to increase by an average of 288 mm (32%) for the 3-year period immediately after cutting and while herbicides were being applied. With the ending of herbicide applications and with the occurrence of natural regeneration, the water yield increases rapidly diminished. Statistically significant but progressively smaller increases occurred in years 4 and 5 (table 13.2). Increases in streamflow were indicated in years 6 through 12, but only those in years 9 and 12 were statistically significant. Beginning in year 13 and continuing through year 34, all changes in annual streamflow were indicated to be decreases. Based only on decreases that were statistically significant (12 of 18 years), streamflow during much of the period of stand regeneration averaged 62 mm/yr less (−7%) than if CF had not been performed.

Water yield increases for the SC, which was cut in thirds over a 4-year period, were rather modest. For years 2 through 4 during, and years 5 through 7 following the SC, there were significant increases in annual water yield ranging from 4% to 9% (table 13.2). As with the CF, regeneration caused water yield to decrease by 3% to 9% compared to the control watershed. And just as with the CF, these decreases persisted for several decades (table 13.2). These decreases are the result of lower stomatal resistance and greater water use by the pioneer species that dominate the early regeneration (Hornbeck et al. 1997b).

Water yield increased 23% the first year following the WT. There were also significant increases of 5% to 8% during years 7, 8, and 13 following cutting. Since year 14 after WT there has been a mixture of small increases and decreases in streamflow (table 13.2). Several possible explanations for the different response on the WT compared to the CF and SC include a lack of regeneration on the skid roads, heavy moose browse near the top of the watershed, and a greater proportion of American beech in the regenerating forest. All these factors could reduce transpiration rates without decreasing water yield (Campbell et al. 2007; Hornbeck et al. 1997b).

The water yield increases occurred primarily as augmentation to low flows during the growing season. Complete recharge of soil moisture usually occurs in mid- to late autumn at HBEF, and transpiration-induced increases in streamflow seldom extend into the dormant season. Depending upon antecedent soil moisture, peak-flow rates during the growing season can be increased by up to 60% in the first 1 to 2 years after harvest. Any increases in peakflow rates quickly disappear with the establishment of a new forest (Hornbeck et al. 1997b). Total snowmelt runoff was largely unaffected by treatments, but timing was changed. In the absence of shade provided by branches and stems, snowmelt and snowmelt runoff were advanced by up to 17 days compared with the control watershed.

The amount and duration of water yield changes at HBEF are determined in large part by differences in how the forests were cut and how rapidly new forests regenerated. The harvesting of the SC was spread over 4 years, allowing regeneration to develop on adjacent strips during harvesting and for water yield increases from cut strips to be utilized by the uncut strips. As a result annual water yield increases from the SC were small and short-lived (table 13.2). The WT was harvested in one large block, hence the larger initial increase in annual water yield. However, the regrowing forest reduced the water yield increases from the WT by more than half in the second year after harvest (table 13.2).

Soil Disturbance and Sediment Yield

Logging on the SC and WT adhered to best management practices (BMPs) prescribed by the state of New Hampshire, but considerable soil disturbance still occurred. For example, surveys after the SC and WT at HBEF showed that 70% and 67% of the respective watershed areas had soil disturbances of varying degrees. Disturbance on the WT (table 13.3) ranged from nearly 4% of the entire watershed having the forest floor intact but depressed by one pass of logging equipment, to 18% covered with wheel or track ruts into mineral soil. Logging disturbed the forest-floor organic horizons to the point where nearly 28% of the WT exhibited bare mineral soil (including scalped mineral mounds, mineral ruts, and bare rocks; Martin and Hornbeck 1994).

The soil disturbances during logging led to some increases in sediment yield. Annual sediment yields for several decades from control watersheds at HBEF (Watersheds (WSs) 3 and 6) varied widely from 1 to 95 kg/ha (table 13.4). By comparison, annual sediment yields since initiation of the SC in 1970 ranged from 3 to 146 kg/ha. The maximum value occurred during the 1973 water year when the second series of strips were cut. Annual sediment yields since initiation of the WT in 1983 ranged from 3 to 208 kg/ha. Sediment yields from the WT watershed prior to harvest had been as great as 134 kg/ha. Statistically significant increases occurred during the first 3 years after WT, and in year 12 (table 13.4).

When sediment reaches a stream it may cause the water to become turbid. This effect is measured in Jackson turbidimeter units, or JTU, and can be used as an index of the effects of harvesting on water quality (Martin and Hornbeck 1994).

Table 13.2 Changes in annual water yield for treated watersheds.

Year after initial treatment	CF Estimated flow if untreated	CF Change due to treatment		SC Estimated flow if untreated	SC Change due to treatment		WT Estimated flow if untreated	WT Change due to treatment	
	(mm)	(mm)	(%)	(mm)	(mm)	(%)	(mm)	(mm)	(%)
1	851	347*	41	777	22	3	649	151*	23
2	954	278*	29	1032	46*	4	883	48	5
3	919	240*	26	1415	116*	8	806	−15	−2
4	902	201*	22	818	68*	8	743	−12	−2
5	840	146*	17	1263	55*	4	682	4	1
6	787	44*	6	867	81*	9	1019	46	5
7	1059	13	1	973	69*	7	1086	52*	5
8	1467	53	4	885	−15	−2	835	66*	8
9	832	67*	8	755	−31*	−4	860	47	5
10	1305	2	0	795	−26*	−3	879	19	2
11	884	48	5	1144	−18	−2	605	22	4
12	996	64*	6	869	−44*	−5	1348	25	2
13	902	−15	−2	1069	−33*	−3	1150	64*	6
14	764	−13	−2	709	−46*	−7	746	−4	−1
15	807	−33	−4	927	−45*	−5	1006	−36*	−4
16	1179	−42*	−4	857	−67*	−8	1112	9	1
17	885	−70*	−8	779	−36*	−5	675	−3	0
18	1098	−62*	−6	702	−59*	−8	740	28	4
19	715	−64*	−9	1080	−43*	−4	622	−5	−1
20	948	−44*	−5	1159	−63*	−5	1198	−6	−1
21	872	−80*	−9	904	−42*	−5	951	−45*	−5
22	790	−83*	−10	943	−59*	−6	1225	6	0
23	708	−55*	−8	918	−30*	−3	1015	−19	−2
24	1110	−35	−3	624	−37*	−6			
25	1194	−48*	−4	1362	−22	−2			
26	923	−36	−4	1181	−18	−2			
27	964	−79*	−8	768	−66*	−9			
28	938	−71*	−8	1020	−46	−4			
29	625	−54*	−9	1132	−34*	−3			
30	1410	−26	−2	673	4	1			
31	1218	−34	−3	771	−18	−2			
32	778	−39*	−5	652	−18	−3			
33	1047	−31	−3	1203	−1	0			
34	1165	−71*	−6	998	−38*	−4			
35	677	18	3	1263	−11	−1			
36	781	9	1	1010	−21	−2			
37	654	70*	11						
38	1241	1	0						
39	1023	19	2						
40	1306	17	1						
41	1036	5	0						

*Change exceeded 95% confidence interval about the calibration regression.

Table 13.3   Soil disturbance on the whole-tree watershed (WT).

Type of disturbance	%	Standard error
Undisturbed	30	3
Depressed (undisturbed but compressed by equipment)	4	1
Scarified (organic and mineral soils mixed)	13	1
Scalped (organic pad scraped from mineral soil)	1	1
Organic mounds (mounds of organic soil)	13	2
Mineral mounds (mounds of mineral soil)	6	1
Organic ruts (wheel ruts lined with organic soil)	10	1
Mineral ruts (wheel ruts into mineral soil)	18	3
Vegetation (stumps and logging slash)	2	1
Bare rocks (rocks > 10 cm diameter)	3	1
Total		100

Three hundred and twenty-five samples for turbidity were collected from a control watershed and the same number of samples were collected from the SC during both storm and nonstorm periods. No samples exceeded 5 JTU from the control watershed. Nine samples (3%) exceeded 10 JTU (the drinking water standard) from the SC, with 40 JTU being the maximum value measured.

Soil Chemical Status

Effects of forest cutting on soil chemistry have been determined only for the WT. Total soil pools of exchangeable Ca, Mg, and K were unchanged in the first 8 years after cutting (Johnson et al. 1997). Decreases in exchangeable cation concentrations in upper soil horizons (Oa and E horizons) were offset by large increases in the spodic horizons (Bh and Bsl). Soil organic matter is the principal source of cation exchange capacity in soils at HBEF. The cation exchange capacity to organic matter ratio increased by about 25% in spodic horizons for the first 8 years after cutting, suggesting that the cutting altered the charge properties and character of organic matter (Johnson et al. 1997).

The mean N pool for the forest floor was 17% lower 8 years after WT (Johnson 1995). Carbon was preferentially lost from soil organic matter, relative to N, resulting in significant decreases in the C/N and C/organic matter ratios in the soil (Johnson 1995).

Stream Water Nutrients

All three experimental treatments at HBEF caused stream water concentrations of Ca^{+2}, K^+, H^+, and NO_3^- to increase, concentrations of SO_4^{-2} to decrease, and concentrations of other ions to change very little. Responses to the WT were intermediate between those from the CF and SC and can be used to demonstrate effects of cutting at HBEF (figure 13.2).

For the WT, mean monthly Ca^{+2} concentrations increased by as much as 3.5 mg/L by the second year after cutting, then gradually declined through year 5

Table 13.4 Sediment yields from four watersheds at the HBEF and from two untreated control watersheds (WS 3 and WS 6)

Water	WS 3	WS 6	SC	WT	WS 6
year	Control watersheds (kg ha^{-1} yr^{-1})		Treatment watersheds (kg ha^{-1} yr^{-1})		Precip. (mm)
1970[b]	—	42	3[a]	—	1360
1971	—	5	13[a]	—	1329
1972[c]	—	6	27[a]	—	1280
1973	—	95	146[a]	—	1565
1974[d]	—	25	16[a]	—	1888
1975	35	18	67	24	1308
1976	10	15	6	12	1769
1977	29	79	132	134	1402
1978	37	18	65	68	1532
1979	47	25	30	97	1362
1980	64	32	38	89	1194
1981	25	34	36	41	1355
1982	28	10	27	35	1585
1983[e]	11	3	5	14	1410
1984[e]	47	35	52	64	1638
1985	5	4	10	112*	1200
1986	45	17	77	129*	1425
1987	71	52	89	208*	1311
1988	4	1	10	6	1290
1989	19	7	35	15	1234
1990	24	13	37	44	1553
1991	17	14	64	33	1669
1992	8	3	13	7	1422
1993	15	12	87	16	1372
1994	8	13	48	11	1405
1995	5	3	3	3	1156
1996	23	71	125	176*	1805
1997	9	7	8	30	1608
1998	3	2	12	10	1290
Mean	25	23			1439
s.e.	4	5			35
C.V.(%)	78	100			13

* Significant increase (p < 0.05).
[a] Estimated by linear regression with WS 6 as the independent variable.
[b] First set of strips harvested on Watershed 4.
[c] Second set of strips harvested on Watershed 4.
[d] Third set of strips harvested on Watershed 4.
[e] Whole-tree clearcutting on Watershed 5.

(figure 13.2). From years 5 through 14, Ca^{+2} concentrations from the WT have remained elevated by an average of 0.5 mg/L when compared to the control watershed.

Concentrations of K$^+$ increased in the first year after WT to a maximum of 1.5 mg/L greater than the control watershed (figure 13.2). K$^+$ from the WT then gradually declined over years 2 and 3 but elevated levels of K$^+$ have persisted for the period of measurement (figure 13.2).

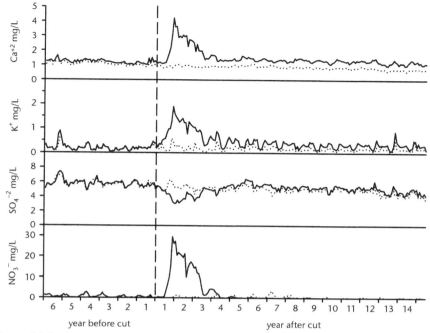

Figure 13.2 Volume-weighted mean monthly concentrations of nutrient ions for the WT (*solid line*) and control watersheds (*dotted line*). The vertical dashed line indicates the beginning of the harvest. Data are from Martin et al. (2000).

Forest harvest has the greatest impact on concentrations of NO_3^-. Concentrations for control watersheds have nearly always been < 5 mg/L, and in recent years have usually been < 1 mg/L (figure 13.2). In the first year after WT mean monthly concentrations of NO_3^- increased from background levels of < 1 mg/L to a maximum of 30 mg/L (figure 13.2). The increases gradually disappeared and concentrations of NO_3^- fell below those for the control watershed by the sixth year after harvest. The decreased level of NO_3^- from the WT has continued through year 14 after harvest.

Concentrations of SO_4^{-2} during precutting ranged between 5 and 7 mg/L and seldom varied between watersheds by more than 0.5 mg/L. WT caused mean monthly SO_4^{-2} to decrease by up to 2 mg/L during the first 3 years after harvest, then return to about the same or slightly higher concentrations than from the control watershed (figure 13.2).

Nutrient Budgets

The changes in water yield and ion concentrations for the harvested watersheds caused streamflow nutrient outputs and net gains or losses from input/output budgets to differ from values for uncut watersheds. Using WT an example, a total of statistically significant increases and decreases in annual streamflow ouputs (table 13.5)

Table 13.5 Input-output budgets in kg/ha/yr for WT. To obtain total output for WT, add output if uncut and change due to cut. Data are from Martin et al. (2000)

Water year after harvest	Ca			K			NO₃-N			SO₄-S		
	Input in precip.	Output if uncut	Change due to cut	Input in precip.	Output if Uncut	Change due to cut	Input in precip.	Output if Uncut	Change due to cut	Input in precip.	Output if Uncut	Change due to cut
1	1.0	7.0	14.0*	0.6	1.5	8.2*	4.5	0.9	28.3*	9.2	11.1	-1.1
2	0.9	8.9	12.3*	0.4	1.7	7.6*	4.2	0.9	28.8*	7.9	14.6	-2.3
3	0.6	8.5	2.9*	0.3	2.0	3.2*	3.6	1.6	3.5*	8.3	12.7	-0.3
4	1.1	7.9	1.9*	0.7	1.4	2.0*	4.6	0.8	-0.3	8.3	12.2	0.9*
5	1.1	7.6	1.8*	0.5	1.8	1.7*	4.8	1.5	-1.3*	8.3	11.4	1.5*
6	1.4	11.1	3.0*	0.8	2.2	2.4*	5.4	3.8	-3.7*	10.1	16.5	3.1*
7	1.2	10.8	4.2*	0.8	1.5	2.5*	5.6	1.9	-1.6*	11.6	17.6	1.8*
8	1.1	8.2	3.8*	0.6	1.4	1.7*	4.8	0.6	-0.4	8.9	13.7	1.9*
9	0.9	7.8	3.2*	0.6	1.4	1.7*	5.0	0.4	-0.3	9.0	13.4	1.2*
10	1.2	8.1	3.1*	0.6	1.4	1.6*	5.8	0.2	-0.2	10.0	4.2	1.0*
11	0.9	6.1	2.1*	0.6	1.0	1.1*	4.8	0.4	-0.4	7.8	9.5	1.2*
12	1.3	10.6	4.9*	0.7	2.3	2.5*	5.8	0.7	-0.6	9.0	19.6	2.6*
13	1.1	8.6	4.5*	0.6	1.8	2.2*	5.3	0.4	-0.2	8.9	15.8	2.2*
14	1.0	6.3	2.0*	0.8	1.4	0.9*	4.4	0.3	-0.2	6.7	10.2	0.9
Sum	14.8	117.5	63.7ᵃ	8.6	22.8	39.3ᵃ	68.6	14.4	54.0ᵃ	124.0	192.5	17.4

*Significant at 0.05 level of probability.
ᵃSum of significant differences only.
Source: Data are from Martin et al. (2000).

shows that in the 14 years since initiation of WT, there has been an increased loss of 64 kg Ca/ha, 39 kg K/ha, 54 kg NO_3-N/ha, and 17 kg S/ha^{-1} (table 13.5).

To put these losses in perspective, streamwater outputs of Ca in the absence of WT would have been 117.5 kg Ca/ha (table 13.5). The additional 63.7 kg Ca/ha lost due to WT thus represents an increase of 54% in Ca outputs in streamwater. Corresponding increases due to stripcutting for other nutrients are 172% for K, 375% for N, and 9% for S.

Despite the large increase in N loss after WT, the watershed still experienced a small net gain in N for the postharvest period due to inputs in bulk precipitation that continued to exceed outputs in streamflow (table 13.5). This was not the case for Ca and K. The input/output budgets for both of these nutrients show substantial net losses before harvesting, and therefore even greater net losses after harvest (table 13.5).

Stream Invertebrates

After the WT, increases in streamflow, light, temperature, and nutrients translated into increased algal abundance with little change in species composition or diversity (Ulrich et al. 1993). The increase in the algal abundance in turn affected the macroinvertebrate community. The standing crop of invertebrates increased the first growing season after cutting due to an increase in herbivorous forms (Burton and Ulrich 1994). At the same time, the populations of two predatory invertebrates also increased. These increases were at the expense of the leaf shredding detritivores.

Comparisons of Results from HBEF with CHL

Forest Regeneration

At CHL, the three communities making up the forest prior to clearcutting had an average basal area of 25.3 m^2/ha (Elliott et al. 1997). The mature, northern hardwood forest on the control watershed at HBEF had an average basal area of 28.3 m^2/ha^{-1} (table 13.1). Hardwood forests at both CHL and HBEF revegetated quickly after the experimental cuttings. However, growth rates during the regeneration period were considerably higher at CHL. By year 16 after the clearcutting at CHL, regeneration for the three major communities found on the watershed had average basal areas between 75% and 105% of that for the preharvest forest (Elliott et al. 1997). In contrast, by year 15 after the WT at HBEF, basal area was 52% of the mature forest (table 13.1). By year 27 on the SC, the basal area of the regenerating forest had reached 86% of that of the mature forest (table 13.1).

At both CHL and HBEF the species composition of the regenerating forest showed some marked differences from the preharvest forest. During the first two or more decades of regrowth, opportunistic species, such as tulip poplar and black locust at CHL and pin cherry and paper birch at HBEF, dominated regeneration (table 13.1) (Elliott et al. 1997). At HBEF, yellow birch, sugar maple, and American beech are gradually increasing on all the treated watersheds and are expected to assume their traditional dominant role between 30 and 40 years after the experimental treatments (Hornbeck and Leak 1992). At CHL some of the oak and hickory species that were dominant in the preharvest forest may not become significant components of the new stand for many decades due to their slow rates of seed dispersal and low survival (Elliott et al. 1997).

Water Yield and Peakflow Rates

The first-year increase in annual water yield after the clearcutting at CHL was 260 mm (Swank et al. 2001), compared to HBEF values of 347 mm after the CF, 116 mm after the second stage of SC, and 151 mm after WT (table 13.2). Increases declined rapidly at both CHL and HBEF, and annual water yields returned within a few mm of precutting levels within 5 to 6 years after conclusion of treatments. However, in the years since the initial recovery, there have been decreases in water yield at both CHL and HBEF. The persistent decreases in annual water yield at HBEF (since year 8 on the SC and year 12 on the CF) are due to the regeneration having substantial numbers of pioneer species that have lower stomatal resistances and greater transpiration than mature northern hardwood forests (Hornbeck et al. 1997b). As the short-lived pioneer species drop from the stand, transpiration is expected to decrease and streamflow should eventually return to precutting levels. An extended period of decreases in water yield has not occurred after WT. Possible explanations are greater area of watershed in skid trails that were slow to regenerate, reductions in regeneration due to browse by moose, and prolific sprouting of beech and sugar maple, which have greater stomatal resistances than pioneer species (Hornbeck et al. 1997b). At CHL the decreases in water yield after clearcutting did not begin to occur consistently until 1994 or 16 to 17 years after harvest. It is likely that these decreases at CHL are also linked to changes in stomatal resistance or leaf area as the regenerating forest goes through successional changes (Swank et al. 2001).

On a monthly basis, the proportionally largest increases occurred during the low flow months of July through October at both CHL and HBEF. However, increases at CHL occurred in nearly every month while at HBEF the increases were restricted to growing season months. At HBEF there is complete recharge of soil moisture on both treated and control watersheds by the start of the dormant season, thus eliminating opportunities for any yield increases until the beginning of the next growing season.

Peakflow rates at CHL increased by an average of 15% during the first 4 years after clearcutting (Swank et al. 2001). The corresponding value for the WT at HBEF was 29% (Hornbeck et al. 1997b), although increases in peakflow rates were as high as 60% in the first 2 years after WT. The increases in peakflow rates diminished quickly at both CHL and HBEF as regeneration became established and created greater soil water deficits.

Sediment Yield

Sediment yields at CHL are generally higher than at HBEF, most likely because HBEF receives less precipitation, has less steep slopes, and has more stony and coarse-textured soils with higher infiltration capacities. In the 2 years before the clearcutting at CHL, annual sediment yields were 230 and 135 kg/ha (Swank et al. 2001). In contrast, annual sediment yields from control watersheds at HBEF averaged about 25 kg ha/yr over the period 1970 to 1998, with a maximum of 95 kg ha/yr (table 13.4). Clearcutting at CHL caused elevated annual sediment yields during roadbuilding, logging, and for a lengthy period after. During the 5- to 15-year period after clearcutting, sediment yields averaged about 340 kg/ha/yr or nearly 50% above pretreatment levels (Swank et al. 2001). Harvesting at HBEF also resulted in increased sediment with maximum values of 146 kg ha/yr during cutting of the second set of strips on the SC, and 208 kg ha/yr during the third year after performing the WT (table 13.4). The increases in annual sediment yield have moderated more quickly after harvest at HBEF than at CHL (table 13.4; Swank et al. 2001).

Sediment yields at both CHL and HBEF are highly variable from year to year and are not correlated with annual precipitation amounts. Instead, sediment yields are driven by the occurrence of large, individual storms, by site differences, and by the specifics of the particular logging operation producing the sediment (Martin and Hornbeck 1994; Swank et al. 2001).

Soil Chemical Status

Effects of cutting on exchangeable soil cations in upper soil horizons were in opposite directions: an increase at CHL (Knoepp and Swank 1996), and a decrease at HBEF (Johnson et al. 1997). At CHL exchangeable Mg and K remained above pretreatment levels at 17 to 20 years after harvest. The decreases in exchangeable cations in upper horizons at HBEF were short-lived (3 to 8 years) and were countered by increases in deeper horizons, with the end result being no net change in exchangeable cations. Total soil N and C concentrations increased in the upper horizons (0 to 10 cm) at CHL by 50% or more in the first 3 years after harvest and remained near or above preharvest levels for 18 years (Knoepp and Swank 1997). In contrast, there were no changes in N and C concentrations in the forest floor at the eighth year after harvest at HBEF. However, soil N and C pools in the forest floor at HBEF were decreased by 17% and 27% respectively at the eighth year after harvest due to reductions in mass of the forest floor (Johnson 1995).

Differences in responses of soil chemical status between CHL and HBEF may be the result of logging disturbances and harvest intensity. Compared to the WT at HBEF, the cable yarding technique and removal of sawlogs only at CHL cretated less soil disturbance and left more biomass to decompose and supply nutrients. At HBEF, the steep midsection of the WT, where logging disturbances were greatest, experienced the greatest losses of soil N and C. Nutrient pools in the relatively flat upper elevations were unchanged.

Streamwater Nutrients and Nutrient Budgets

Table 13.6 contrasts nutrient ions in streamflow and bulk precipitation for control watersheds at CHL and HBEF. Precipitation and streamflow are dilute at both locations, but HBEF is more acidic due to higher concentrations of SO_4^{-2} and NO_3^- and minimal buffering by HCO_3^-.

Table 13.6 Volume-weighted mean annual concentrations* of dissolved inorganic concentrations for undisturbed watersheds at HBEF and CHL.

Substance	Bulk precipitation		Streamflow	
	HBEF	CHL	HBEF	CHL
Ca^{+2}	0.093	0.194	1.153	0.583
Mg^{+2}	0.025	0.041	0.305	0.326
K^+	0.049	0.094	0.192	0.499
Na^+	0.088	0.170	0.835	1.220
NH_4^+	0.095	0.183	0.016	0.002
H^+	0.058	0.027	0.011	0.000
SO_4^{-2}	2.169	1.590	5.859	0.450
NO_3^-	1.674	0.143	1.073	0.003
Cl^-	0.271	0.262	0.474	0.662
PO_4^{-3}	0.021	0.013	0.003	0.006
HCO_3^-	—	0.074	1.620	4.970
Si	T**	0.030	4.592	8.800
pH	4.23	4.57	4.96	>6.00

* Data are means for 1973–1983.
** T = trace
Sources: Data from Likens and Bormann (1995); and Swank and Waide (1988).

Streamwater concentrations and nutrient budgets at HBEF are more responsive to cutting disturbances than at CHL. To illustrate, the WT at HBEF caused streamwater concentrations of nutrient ions to increase by maximums of 3.5 mg/L for Ca^{+2}, 1.5 mg/L for K^+, and 30 mg/L for NO_3^- Corresponding maximum increases after clearcutting at CHL were 0.4 mg/L for Ca^{+2}, 0.5 mg/L for K^+, and 0.7 mg/L for NO_3^-. The increases in streamwater concentrations from the treated watersheds translated into increased nutrient losses for the first six years after harvest of 35.9 kg Ca/ha at HBEF versus 12.0 kg Ca/ha at CHL; 25.1 kg K/ha for HBEF versus 8.4 kg K/ha at CHL; and 55.3 kg NO_3-N/ha at HBEF versus 3.8 kg NO_3-N/ha at CHL (Swank et al. 2001, Martin et al. 2000). Swank et al. (2001) attributed the resilience of nutrient cycles to disturbance at CHL to large pools of organic matter and elements that turn over slowly and to the high rates of net primary productivity and sequestration and storage of nutrients in successional vegetation.

An interesting difference between CHL and HBEF occurred in the nitrogen cycle during regeneration. At the fourth year after both the SC and WT, losses of NO_3-N became less than from mature forests and continued as such for a decade or more (table 13.5). This pattern has been attributed to uptake and sequestration by the regrowing forest (Martin et al. 2000). At CHL, there was a second and more sustained pulse of NO_3-N that began around the 15th year after clearcutting. This pulse, which has not occurred at HBEF, has been attributed to a series of events including reduction in uptake due to mortality of early successional species (including black locust, an important N-fixing species), nutrient release from woody decomposition, elevated soil nitrogen transformations, and reduction in soil C/N ratio (Swank et al. 2001).

In general, the increased nutrient losses via leaching to streamflow are relatively small after harvests at both CHL and HBEF and should not impact productivity in the next rotation. The increased losses represent minute portions of total site nutrient capitals and are small relative to nutrients removed in biomass. The only possible concern might be with losses of base cations from soils at HBEF, where capitals are significantly lower than at CHL.

Stream Invertebrates

At HBEF, WT reduced the species diversity of stream invertebrates, but increased the abundance (Burton and Ulrich 1994). At CHL, clearcutting was accompanied by a greater sediment load than at HBEF and impacted all aspects of the invertebrate habitat and community. However, by 16 years after clearcutting, benthic invertebrate abundance was 3 times higher and invertebrate biomass and production were two times higher than in an adjacent control stream (see Wallace and Ely, chapter 11, this volume).

Conclusions

Despite significant differences in site characteristics between CHL and HBEF, responses to intensive harvests showed several similarities:

- Harvested sites regenerated rapidly, with opportunistic and pioneer species dominating regrowth for the first 20+ years after harvest.
- Water yield increases occurred during the early years after harvest but declined rapidly with regrowth. Changes in species composition eventually resulted in decreases in water yield when compared to mature forests. Water yield increases were proportionally largest during late summer and early autumn. Peakflow rates were increased by 30% to 60% immediately after harvest.
- Sediment yields increased at both locations but were minimized by careful roading and logging practices.
- Harvesting caused contrasting responses in soil chemical status but in general the harvests at both sites did not cause adverse impacts on soil cations, N, or C.
- Streamwater concentrations of Ca^{+2}, K^+, and NO_3^- and their corresponding output budgets were increased after harvest, but water quality was not adversely impacted and losses from nutrient capitols were relatively small and should not impact site productivity.

Results from both CHL and HBEF show that intensive harvests can be conducted with minimal impact on hydrologic and nutrient cycles and sediment yields. Careful planning of harvesting operations and application of BMP are imperative to achieving these results. It is important to realize that species composition of the regrowing forest will, at least initially, be dramatically changed from that of the previously harvested forest.

Literature Cited

Burton, T. M., and K. E. Ulrich. 1994. The effects of whole-tree harvest on insects associated with leaf packs in small streams in New Hampshire. *Verhandlung der Internationalen Vereinigung für Theoretische und Angewandte Limnologie* 25: 1483–1491.

Campbell, J. L., C. T. Driscoll, C. Eagar, G. E. Likens, T. G. Siccama, C. E. Johnson, T. J. Fahey, S. P. Hamburg, R. T. Holmes, A. S. Bailey, and D. C. Buso. 2007. Long-term trends from ecosystem research at the Hubbard Brook Experimental Forest. General Technical Report NRS-17. U.S. Department of Agriculture, Forest Service, Northern Research Station. Newtown Square, Pennsylvania.

Elliott, K. J., L. R. Boring, W. T. Swank, and B. R. Haines. 1997. Successional changes in plant species diversity and composition after clearcutting a southern Appalachian watershed. *Forest Ecology and Management* 92: 67–85.

Hornbeck, J. W., S. W. Bailey, D. C. Buso, and J. B. Shanley. 1997a. Streamwater chemistry and nutrient budgets for forested watersheds in New England: variability and management implications. *Forest Ecology and Management* 93: 73–89.

Hornbeck, J. W., and W. B. Leak. 1992. Ecology and management of northern hardwood forests in New England.General Technical Report NE-159. US Department of Agriculture, Forest Service, Northeastern Forest Experiment Station. Radnor, Pennsylvania.

Hornbeck, J. W., C. W. Martin, and C. Eagar. 1997b. Summary of water yield experiments at Hubbard Brook Experimental Forest, New Hampshire. *Canadian Journal of Forest Research* 27: 2043–2052.

Johnson, C. E. 1995. Soil nitrogen status eight years after whole-tree clear-cutting. *Canadian Journal of Forest Research* 25: 1346–1355.

Johnson, C. E., R. B. Romanowicz, and T. G. Siccama. 1997. Conservation of exchangeable cations after clear-cutting of a northern hardwood forest. *Canadian Journal of Forest Research* 27: 859–868.

Knoepp, J. D., and W. T. Swank. 1996. Long-term effects of commercial sawlog harvest on soil cation concentrations. *Forest Ecology and Management* 93: 1–7.

Knoepp, J. D., and W. T. Swank. 1997. Forest management effects on surface soil carbon and nitrogen. *Soil Science Society of America Journal* 61: 928–935.

Likens, G. E., and F. H. Bormann. 1995. *Biogeochemistry of a Forested Ecosystem.* 2nded. Springer-Verlag, New York, New York.

Martin, C. W., and J. W. Hornbeck. 1994. Logging in New England need not cause sedimentation of streams. *Northern Journal of Applied Forestry* 11: 17–23.

Martin, C. W., J. W. Hornbeck, G. E. Likens, and D. C. Buso. 2000. Impacts of intensive harvesting on hydrology and nutrient dynamics of northern hardwood forests. *Canadian Journal of Fisheries and Aquatic Sciences* 57: 19–29.

Swank, W. T., J. M. Vose, and K. J. Elliott. 2001. Long-term hydrologic and water quality responses following commercial clearcutting of mixed hardwoods on a southern Appalachian catchment. *Forest Ecology and Management* 143: 163–178.

Swank, W. T., and J. B. Waide. 1988. Characterization of baseline precipitation and stream chemistry and nutrient budgets for control watersheds. Pages 57–79 in *Forest Hydrology and Ecology at Coweeta*. W. T. Swank and D. A. Crossley, Jr., editors. Springer-Verlag, New York, New York.

Ulrich, K. E., T. M. Burton, and M. P. Oemke. 1993. The effects of whole-tree harvest on epilithic algal communities in headwater streams. *Journal of Freshwater Ecology* 8: 83–92

14

Bridging the Gap between Ecosystem Theory and Forest Watershed Management

A Synthesis of 30+ Years of Research on WS 7

Jackson R. Webster*
Wayne T. Swank
James M. Vose
Jennifer D. Knoepp
Katherine J. Elliott

Introduction

The history of forests and logging in North America provides a backdrop for our study of Watershed (WS) 7. Prior to European settlement, potentially commercial forests covered approximately 45% of North America (Clawson 1979), but not all of it was the pristine, ancient forest that some have imagined. Prior to 1492, Native Americans had extensive settlements throughout eastern North America (Mann 2005), but to European settlers, the area was a wilderness. It was described by early settlers as "repugnant, forbidding, and repulsive ... The forests were wild areas, alien to man and in need of felling, firing, grazing, and cultivating so that they could become civilized abodes" (Williams 1989). Across North America, forests were cleared for agriculture and forest products, primarily lumber and fuel (MacCleery 1992). First in the Northeast, then in the Midwest, the Great Lakes region, the Southeast, and the Pacific Northwest, forests were cleared, with little regard for future forest values. Forests were viewed as an inexhaustible natural resource, and large logging companies would "cut and run" to the next tract of forest. By the mid-nineteenth century, commercial forest land in the United States had been reduced to about half its original area. In the southern Appalachian region, almost 90% of the forests were cut; and many of these

* Corresponding author: Department of Biological Sciences, Virginia Polytechnic Institute and State University, Blacksburg, VA 24061 USA

areas were burned by the turn of the century (Yarnell 1998). In the later nineteenth century, George Perkins Marsh, Frederic Starr, and others began to raise concerns about extensive forest loss. Scientists such as Bernhard Fernow and Gilford Pinchot began the era of forest management in the United States. As a result of improved forest management, declining demand for forest products (especially fuel), fire suppression, and agricultural land abandonment, the area of forest land began to increase (Clawson 1979). "Regrowth can be seen everywhere, and one is struck by the robustness of the forest" (Williams 1989). The resilience of American forests is especially evident in the southern Appalachians. For example, there was a 38% increase in wood volume in the forests of the southern Appalachian region of North Carolina between 1984 and 2006, with no change in forest area (Fox et al. 2010).

When the original research proposal was developed in 1974 to examine ecosystem response to logging, the ideas were based on fairly simple concepts of stability (Monk et al. 1977) and on two aspects of ecosystem response to disturbance, resistance and resilience (Webster et al. 1975). First, ecosystems exhibit varying degrees of resistance to disturbance—stable systems show little change in response to a disturbance; whereas less-stable systems show large change. Second, resilience, the ability of an ecosystem to recover following disturbance, also differs: resilient systems recover rapidly; nonresilient systems are much slower to recover. These concepts were consistent with ecosystem theory 35 years ago (e.g., Holling 1973). Ecosystem stability concepts and lexicon have broadened extensively since then (e.g., Waide 1988; Grimm and Wissel 1997; Hooper et al. 2005); however, these two fundamental concepts remain central to any discussion of ecosystem stability (e.g., Levin and Lubchenco 2008). Our research on the response of the Coweeta WS 7 to logging provides an opportunity to intensively examine ecosystem response to disturbance from the evolving perspective of ecosystem stability.

Our original notions of ecosystem stability were derived from engineering concepts in which disturbance is an impulse function that can be represented by an instantaneous change in initial conditions (Waide and Webster 1976). Many ecological disturbances, such as fire and flash floods, approximate instantaneous change; other disturbances, however, occur over longer periods of time, even when related to the recovery time of ecosystems. These two types of disturbances have been termed *pulse* and *press* disturbances, respectively (Bender et al. 1984). Early in the studies of WS 7, it was recognized that logging the terrestrial system approximated a pulse disturbance; whereas, the stream on WS 7, Big Hurricane Branch, was responding to a press disturbance (Webster and Patten 1979; Gurtz et al. 1980; Webster et al. 1992). This press disturbance results from the continued modification of organic (leaf and wood) inputs and nutrient supply to Big Hurricane Branch from the terrestrial system (see Swank et al., chapter 3, and Webster et al., chapter 10, this volume).

Assumption of Asymptotic Stability

The original research proposal also made the assumption that ecosystems are asymptotically stable; that is, that following a small disturbance, ecosystems will return to their original state. However, ecosystems are not globally stable—more extreme disturbances may move an ecosystem outside its domain of stability (sensu

Holling 1973; Gunderson 2000) to an alternate state. This has occurred in forest ecosystems in eastern United States; for example, with the loss of a foundation species, such as American chestnut (Ellison et al. 2005; Elliott and Swank 2008), and extreme disturbances, such as the deposition of toxic chemicals from copper smelting at Copper Hill, Tennessee, and the massive erosion during cotton farming, which resulted in loss of much of the organic soil from the southern United States. (Richter and Markewitz 2001). These extreme disturbances may result in permanently altered systems. Significant management intervention may restore some system attributes, for example, the planting of nearly 20 million trees and shrubs at Copper Hill may eventually result in an ecosystem that is similar in structure to what was there before. European settlement (Johnson 2002). However, it may take centuries for full recovery of ecosystem structure and function. In the original proposal, we hypothesized that the disturbance impacts of commercial sawlog harvest would not move a watershed ecosystem to an alternate stable state.

What Was the Original Condition of WS 7?

Interpreting ecosystem responses to contemporary disturbances must be viewed in the context of historical disturbance regimes because in many cases, disturbance responses are shaped by the legacy of earlier disturbances. WS 7 was certainly subjected to many disturbances: a major hurricane in 1835, burning by Native Americans prior to 1837, logging and grazing in the nineteenth century, and chestnut blight in the early twentieth century (see Boring et al., chapter 2, this volume)

By 1975, WS 7 was clearly not a pristine, old-growth forest due to a wide range of human and natural disturbances. It is likely that these historical disturbances have influenced the postharvest response trajectory on WS 7. The time required for forest recovery to a preharvest condition in terms of tree-species composition, age–class distribution, and soil chemistry would be substantially longer than the time required for forest maturation defined as when the forest reaches an age at which harvesting is economically viable. For example, in the case of nutrient cycling, Swank (1984) showed that annual nutrient inputs in bulk precipitation at Coweeta can exceed nutrient removal associated with sawlog harvest over a typical rotation period (70–80 yr). However, other research at Coweeta using simulation modeling of nitrogen cycling (Waide and Swank 1977) showed that whole tree harvest repeated for shorter rotations (25–30 yr) leads to long-term reductions in total site nitrogen and nutrient availability Our studies on WS 7 and similar studies at the Hubbard Brook Experimental Forest in New Hampshire suggest that when best forest management practices are used, sufficient residual material is left on site to replenish soil nutrients and that there is little site degradation in terms of nutrient capital loss (see Hornbeck et al., chapter 13, this volume). On the other hand, the time for recovery to replenish large wood in streams would need to be measured in centuries rather than years (Webster et al. 1992).

What Is the Trajectory of Response?

Disturbances during the past 30 years, for example, changes in atmospheric chemistry and the decline or loss of tree species (see Swank and Webster, chapter 1, this volume), have influenced the response at WS 7 as well as altered the reference forests and streams at Coweeta. Tree-species-composition changes that have occurred include pitch pine decline due to drought and concurrent southern pine beetle infestation (Elliott and Vose 2005; Nowak et al. 2008) and eastern hemlock mortality due to hemlock woolly adelgid (Ellison et al. 2005; Elliott and Vose 2011). The combined effect of natural disturbances such as disease, insects, drought, and fire along with the human-caused disturbances on ecosystem processes mean that the recovery trajectory of WS 7 is proceeding toward a continuously changing target. The ability of an ecosystem to respond in this kind of changing-stability landscape has been referred to as its *adaptive capacity* (Gunderson 2000).

Chemical changes in Big Hurricane Branch also show trends that reflect changing reference conditions. Meyer et al. (chapter 6, this volume) found that the dissolved organic carbon in Big Hurricane Branch followed the expected return toward reference levels for 7 years after the clearcut, but this trend did not continue for the next 20 years due to long-term trends in both WS 7 and the reference steam. Worrall et al. (2003) used time series analysis to examine long-term stream nitrate concentrations on WS 7 and WS 2 from 1971 to 1997. Auto-regressive modeling revealed a significant annual memory effect in both watersheds, but WS 2 responded to drought conditions and WS 7 responded primarily to vegetation changes. Moreover, on WS 7 a significant impulse function was derived for nitrate export in 1989–1997, suggesting the watershed is in a temporary equilibrium (Worrall et al. 2003). This impulse response was again observed in the stream nitrate dynamics in 2002–2007 (see Swank et al., chapter 3, this volume). Worrell et al. (2003) concluded that clearcutting has modified watershed nitrogen dynamics beyond the limit of stability and that reversal of this trend would require "massive management intervention." The long-term NO_3-N budget (see Swank et al., chapter 3, this volume) also continues to provide strong evidence that WS 7 is still in latter stage 1 of watershed N saturation (sensu Aber et al. 1989) as described for earlier years of forest succession (Swank and Vose 1997).

Stability of What?

Is ecosystem resilience based on the return of species composition or on some ecosystem function? Clearly, the two are closely tied, since ecosystem response depends on the functional characteristics of the species involved. In the forest, the species shift from oaks and hickories to tulip poplar, red maple, and black locust has resulted in faster decomposition and higher nutrient content in the litter layer. However, the greater abundance of ericaceous shrubs may moderate this change in litter quality and may also inhibit the regrowth of some late successional species (see Boring et al., chapter 2, this volume). In addition, the shift in species composition has increased transpiration, with a consequent decrease in streamflow in some years (see Swank et al., chapter 3, this volume). Where a single species has a unique functional trait, its abundance may fundamentally affect ecosystem response. For example, black locust (*Robinia pseudoacacia*) is the most important tree species with nitrogen-fixing symbionts in the forests of eastern United States. Its abundance in the first 10–20 years following logging contributes to the long-term elevation of stream nitrate concentrations (see Swank et al., chapter 3, this volume).

A similar interplay of the resilience of species composition and ecosystem function has been observed in Big Hurricane Branch (see Webster et al., chapter 10, and Wallace and Ely, chapter 11, this volume). Shifts in the structure and function of benthic invertebrates on WS 7 have been an integrated product of alterations in the physical habitat (e.g., sediment), food base, nutrient dynamics, light levels, and the temperature of the stream. Immediately after clearcutting, macroinvertebrate taxonomic diversity increased in Big Hurricane Branch and was accompanied by a shift in functional benthic groups to those that feed on algae, that is, scrapers and collector-gatherers. Gurtz and Wallace (1984) found that many taxa decreased in abundance in areas of lower stream gradient (sand and pebble habitat); whereas taxa increases were observed in the steepgradient bedrock-moss habitat. Following the rapid growth of vegetation on WS 7, allochthonous litter input to Big Hurricane Branch returned to near pre-logging levels by 1983 (Webster et al. 1992). Consequently, detritus-feeding benthic macroinvertebrates (i.e., shredders) responded rapidly, and 9–10 years after logging their production was greater than that in a reference stream, which was attributed to a greater abundance of early successional leaf litter (Stout et al. 1993). Following 16 years of succession, benthic macroinvertebrate abundance was still 3 times higher, and macroinvertebrate biomass and production were 2 times higher (habitat weighted) in Big Hurricane Branch compared to the reference stream (Stone and Wallace 1998). However, by 2003, abundance, biomass, and production of most macroinvertebrate groups were similar to the reference stream (Ely and Wallace 2010). This apparent recovery of the macroinvertebrate assemblage occurred even though benthic organic matter in the WS 7 stream was still only about half that of the reference stream (Ely and Wallace 2010; see also Wallace and Ely, chapter 11, and Webster et al., chapter 10, this volume). It is evident that despite the similarities in benthic macroinvertebrate structure and function, logging activity continued to influence this group of consumers even 25 years after logging, probably because of enhanced resource quality (Ely and Wallace 2010).

Watershed logging has a multifaceted effect on stream macroinvertebrate assemblages. Probably the most important factors are changes in resources and in sediment, but it is difficult to separate the effects of these two factors. Most interpretations have been based on changes in resources. However, the different responses of the macroinvertebrates in different habitats suggest a possibly important role of sediment. Recovery was most rapid in the more stable, moss-covered bedrock habitat. The slower recovery in the riffle and depositional habitats (see Wallace and Ely, chapter 11, this volume) might be attributable to the negative effects of fine sediment resulting from road building and logging.

The Need for a Long-Term Perspective

The need for a long-term perspective of forest ecosystem response to disturbance is clear. Ecosystems that appear to be in rapid recovery from multiple disturbances may be masking longer-term shifts in ecosystem behavior (Palumbi et al. 2008). For example, the pattern of nitrate export from WS 7 during the first several years following logging appeared to be exactly what we expected—a rapid increase in nitrate export followed by an asymptotic return approaching pre-logging levels (see Swank et al., chapter 3, this volume). However, beginning around 1990, 13 years after harvest, NO_3 concentrations began rising again, reaching peak concentrations in 1996–1997 that were about 35% above earlier post-logging levels. Subsequently, NO_3 concentrations again declined toward pre-logging levels, showing some short-term elevated concentrations in 2003–2005.

Similarly, the initial water yield increase due to decreased evapotranspiration after cutting and subsequent asymptotic return to baseline flows in the ensuing 10 years were as expected from decades of hydrologic research at Coweeta. However, about 17 years after harvest, annual water yield frequently declined below pre-logging levels, suggesting increased evapotranspiration in WS 7. Possible explanations for the long-term trends in stream NO_3 and water yield are given by Swank et al. in chapter 3 of this volume. The long-term studies of invertebrates in Big Hurricane Branch also illustrate a rapid recovery from initial changes followed by a much longer period of continued change (see Wallace and Ely, chapter 11, this volume). These multidecade responses clearly illustrate significant legacies of the logging disturbance and the need for long-term studies to detect and determine cause-and-effect relationships.

Management Implications

Findings from WS 7 provide important information on the management of southern Appalachians mixed-hardwood forests. At the time the study was initiated, conventional logging was typically conducted by tractor yarding from closely spaced skid roads. In contrast, the high-lead cable logging operation on WS 7 used a 2-drum yarder with a 9-m boom mounted on a truck, with a mainline of 320 m of wire cable and 915 m of haul back line (see figure 1.3 in chapter 1). Operating from a road, the yarder had the capacity to yard whole tree logs a distance of up to 250 m and to suspend them above the ground. Thus, there was minimal forest floor and soil disturbance. Conventional logging would have required more than twice the miles of logging roads that the high-lead system required.

A detailed economic analysis was conducted using data collected on the direct costs per unit of wood volume for cable yarding compared to a conventional logging system on the same area (Robinson and Fisher 1982). From an economic perspective, the cable system was competitive with conventional logging; moreover, the reduced environmental impact on soil and water resources from cable yarding clearly favored this method for logging on steep slopes. Within 2 years, timber sales on National Forest lands typically required cable logging on slopes exceeding 35%. In subsequent years, the technology and expansion of cable logging have advanced and been modified to meet harvesting requirements for a wide range of silviculture prescriptions, and today about 40% of Forest Service timber sales in this region require cable logging.

The WS 7 study also provided an opportunity to conduct detailed research on the effectiveness of previously established best management practices for forest road construction activities and to evaluate some new road standards meant to reduce erosion and sediment movement (Swift 1988). New information useful to managers included an assessment of the portion of total soil loss that comes from cut slopes, road beds, or fill slopes and the effectiveness of grass and gravel in reducing soil loss. The road research also evaluated filter strip standards downslope from roads, providing techniques for controlling soil loss.

In chapter 3 of this volume, Swank et al. presented sediment yield data showing that roads were the major source of sediment delivered to the streams on WS 7. We also emphasize that the erosion response was due to the storms in May 1976. Because of the record rainfall and discharge during this event, which occurred before the new roads were stabilized, we suggest that the magnitude and duration of sediment yield measured on WS 7 are in the upper limits that could be expected in the region after clearcutting and cable logging.

This unique erosion event and set of conditions provides an opportunity to evaluate the long-term effects of such a major disturbance on stream structure and function. Following the initial pulse of sediment export from the catchment, there was a continued release, over a 15-year period, of sediment from upstream storage that had been primarily deposited during the 1976 storms (see Swank et al., chapter 3, this volume). Concurrent with stream sediment dynamics, stream faunal studies indicated that this storm was a major determinant of macroinvertebrate response and recovery on WS 7 (see Wallace and Ely, chapter 11, this volume).

Water yield responses measured following harvest on south-facing WS 7 support the regional use of a previously derived empirical model for predicting short-term annual flow responses for forest management planning in similar hardwood forests. Water yield increases were distributed throughout the year, with the longest percentage flow increases occurring in the autumn when flows are normally lowest and water demands are high. The effects of the forest management prescription on stormflow parameters was low due to inherent hydrologic factors of the watershed, the low density and proper design of roads, and mineral soil disturbance associated with cable logging. The small initial nutrient losses following logging provide evidence for minimal impact of the management prescription on ecosystem health. High rates of net primary production and storage of nutrients in successional vegetation were mainly responsible for nutrient retention.

Summary: Perceptions of Long-Term Changes in Economic Values and Ecological Services in a Management Context

In table 14.1 we have attempted to summarize the status of watershed parameters before systematic observations began; the changes the first few years after logging; current conditions; and changes we predict for the next 65 years. Based on this summary, figure 14.1 illustrates changes to economic values (primarily extractive timber values) and ecological services that have occurred. It also shows predicted changes both with and without the influence of human caused environmental changes (such as climate change, invasive diseases, anthropogenic nutrient deposition). Prior to European settlement, the forest was a highly valuable resource with high timber value and clean water (State 1 in figure 14.1 and column 1 in table 14.1). Douglass and Hoover (1988) quoted Mr. C. E. Marshall in this description of the Coweeta area in the early twentieth century: "For reproduction of desirable hardwoods, there are no better lands in Western North Carolina." WS 7 also had high ecological value, with attributes such as highly diverse assemblages of trees, wildlife, and stream fauna. It also provided many services, which we recognize today as ecological and economic services, including high quality water and sediment retention. However, the watershed was already impacted by humans; for example, the relatively open understory in the southern Appalachians described by early explorers was probably the result of intentional burning by Native Americans (Mann 2005), so it is likely that WS 7 was frequently burned. Prior to our logging experiment, the watershed was already somewhat degraded from pre-European settlement conditions by fire suppression, selective logging, woodland grazing (by early settlers and experimentally in the 1950s), chestnut death, and perhaps other species losses (State 2 in figure 14.1 and column 2 in table 14.1). We cannot quantify the effects of the loss of formerly abundant species, such as passenger pigeons, and potential keystone predators, such as wolves and mountain lions during this period; however, we know that many ecosystems shifts in predator-prey relationships have a substantial effect on ecosystem, processes. For example, at both Fernow and Hubbard Brook, large animals (deer and moose, respectively) have altered the forest recovery from cutting (see Adams and Kochendenfer, chapter 12, and Hornbeck et al., chapter 13, this volume).

Without the experimental logging and further environmental changes, the system would probably have moved by natural successional processes to a new stable state with higher economic value and different ecological values (figure 14.1, State 3). Perhaps, from a Clementsian viewpoint (Clements 1916, 1936) and a Thoreau system of values (Thoreau 1860; see Foster 1999), there is a state on the right side of the figure that ecosystem function and structure might tend toward, with constant environmental conditions and no human influences (State 0). Because environmental conditions are not constant, State 0 is purely conceptual—pre-European climates were not constant; drought periods and hurricanes happened. Indeed, the fact that our forests respond so quickly to disturbance suggest a long history of disturbances and environmental variation. If State 0 did exist, we might not judge it to have extremely high economic timber value by today's evaluation standards because it would likely have contained considerable low-quality timber due to insects and disease.

In its current state (column 4 in table 14.1), WS 7 has clearly not reached most of the structural and functional characteristics of pre-European settlement ecosystems, but it may be too soon to tell if it is on that trajectory since this forest is only 37 years old. Indeed, we hypothesize that the ultimate stable state of WS 7 (without further logging, State 5 in figure 14.1) will be considerably different than that of the pre-European settlement state because of the loss of American chestnut and other irreversible changes, such as new climate and disturbance regimes. Chapin and Starfield (1997) used the term "novel ecosystems" to refer

Table 14.1. Characteristics of watersheds before the European settlement (ca. 1700), just prior to logging WS 7 (1975), the first few years after logging WS 7, at the present time (37 years after logging WS 7), and predicted after 100 years of succession with current management practices and anticipated environmental changes.* Interpretations of the magnitude of parameters (i.e., low, moderate, high) are relative across the five time periods.

Parameter	Before European settlement (ca. 1700); State 1 in fig.14.1	Reference (WS 7 before road building and logging, 1974 (WS 2, WS 14); State 2 in fig. 14.1	WS 7 shortly after logging (1976-1980); State 4 in fig. 14.1	WS 7 now, 37 yr after logging (2010)	WS 7 after 100 yr of succession (2075); State 5 in fig. 14.1
Stand characteristics	Large, old trees	Mature trees	Rapid regrowth, saplings, small trees	Black locust being replaced by tulip poplar	Dominated by large tulip poplar
Tree distribution	Patchy	More uniform	Open, mid-story	Fairly uniform	Increasingly patchy
Tree composition	American Chestnut/red oak/hickory	red oak/hickory	black locust/tulip poplar/red maple	tulip poplar/red maple/chestnut oak	tulip poplar/red maple/chestnut oak
Woody diversity	Moderate	Slightly reduced	Decreased	Unchanged	Moderate
Herbacious diversity	Low	Moderate	Increasing	Decreasing	Moderate
Net primary production	Low	Moderate		Decreasing	
Vegetation biomass	High, 384 t/ha (Boring et al., ch. 2)	Moderate, 156 t/ha (Boring et al., ch. 2)	Low, 1.4 to 25 t/ha (Boring et al., ch. 2)	Aggrading, > 88 t/ha (Boring et al., ch. 2)	High, approaching old-growth (Vose and Bolstad 2007)
Forest floor biomass	High, 27 t/ha (Vose and Bolstad 2007)	High, 26 t/ha (Vose and Bolstad 2007)	High, decreasing	Decreasing to moderate	Moderate
Soil organic C and N	Very high	High	Very high	Decreasing	Moderate
Soil NO$_3$	Low	Low	High	Still high	
Streamwater NO$_3$	Moderate	Low (> 10 μgN/L)	Very high (100 μgN/L)	Still high (70 μgN/L)	
Sediment load	Very low	Very low	High	Low, except storms	Low
Evapotranspiration	Low	Moderate	Low	High	Moderate to Low
Peakflows	Low	Low	Increase avg. of 15%	Low	Low
Stormflow volume	Low	Low	Increase avg. of 10%	Low	Low
Canopy insects	Diverse	Diverse	Responding to vegetation changes	Low diversity	Low diversity
Stream invertebrates	Diverse, detritus based	Diverse, detritus based	Low diversity, grazers	Detritus based, reduced production	Detritus based, reduced production
Stream benthic organic matter	High, refractory	High, refractory	Low	Low, more labile	Low, more labile
Stream large wood	Very high	High	High (slash), except low where removed	Low	Still low

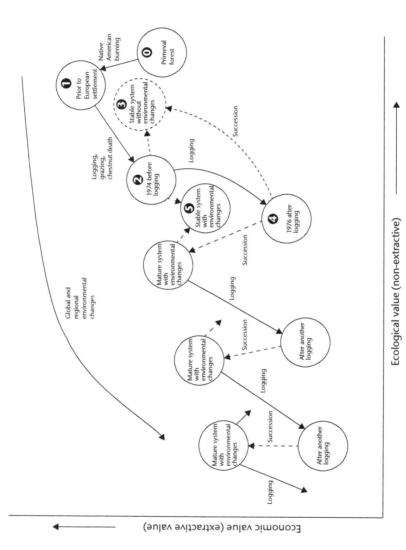

Figure 14.1 Hypothetical trajectories of change in ecological and extractive values of WS 7. Solid lines are human-caused changes; dotted lines represent natural successional changes occurring without global and regional environmental changes; and dashed lines are natural successional changes occurring with global and regional environmental changes. These global and regional environmental changes are predicted to cause decline in both economic and ecological values.

to ecosystems responding to global climate change, but Hobbs et al. (2006) and Seastedt et al. (2008) noted that novel ecosystems could result from a broad range of both abiotic and biotic alterations. In that sense, WS 7 in 1974 was already a novel ecosystem.

Immediately after the 1976 clearcut, the watershed had reduced economic value (State 4 in figure 14.1 and column 3 in table 14.1) because there were no harvestable trees remaining. Some ecological service values were likely changed. For example, stream water quality was degraded by sediment even though chemical levels remained well below drinking water standards. There was little reduction in other ecological values, and some values such as improved habitat for early succession-dependent wildlife species and soft mast production were undoubtedly enhanced. Without environmental change, succession would theoretically move the system back towards State 3. In actuality, succession is moving the system towards a new state (State 5 in figure 14.1 and column 5 in table 14.1), though this new state is unknown due to climate change, diseases such as dogwood anthracnose, invasive animals such as earthworms, hemlock woolly adelgid, and perhaps gypsy moths, and invasive plants such as Japanese stiltgrass. Without forward-looking management, we predict that future clearcut logging combined with continuing climate and other environmental changes will reduce both the economic and ecological values of this forested watershed. For example, some evidence can be derived from WS 13 at Coweeta, where a second experimental cutting produced a more simplified stand structure made up mostly of sprout origin *Liriodendron tulipifera* (Elliott and Swank 1994). Also, computer simulations of Coweeta forests suggest loss of nutrient capital and reduction of forest production following multiple cycles of logging (Waide and Swank 1977). The question we pose is whether we can improve both economic and ecological values with appropriate management practices (State 6 in figure 14.2). The answer is clearly yes. The even-aged clearcut prescription for WS 7 has resulted in a forest dominated by *Liriodendron tulipifera* (see Boring et al., chapter 2, this volume). This provides various management options depending on predictions of future timber values. Thinning might be one tool to achieve desired future stand structure and species composition. Proper road construction and logging techniques that minimize soil disturbance cause less deterioration of water quality. With continued road maintenance, future harvest will require very little soil disturbance. In later succession, prescribed fire might be used to encourage more valuable timber species, such as oaks (Arthur et al. 2012). Although we once thought that slash should be removed from channels, we now know that wood in streams is good for sediment retention and animal habitat (e.g., Dolloff and Webster 2000; Gregory et al. 2003). Man-made wood structures placed in streams may further enhance aquatic animal habitat. The construction and maintenance of trails can enhance recreational values.

In managing this ecosystem, we must "not only anticipate change, but we must acknowledge that current systems have already been transformed and are in the process of transforming further" (Seastedt et al. 2008). The species composition and biogeochemical conditions are products not only of current climate change but of other anthropogenic environmental changes, including elevated CO_2, nitrogen deposition, forest diseases, and exotic invasions. We contend that it will not

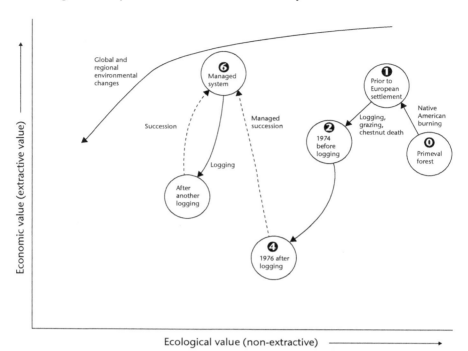

Figure 14.2 Hypothetical trajectories of change in ecological and extractive values of WS 7 with adaptive management of post-logging succession. Solid lines are human-caused changes and dashed lines are successional changes with management based on ecosystem research and predictions of environmental change.

be possible to restore the ecosystem back to some historic or prehistoric state because it has moved beyond the boundaries of the stability of those states. The goal of management cannot be to restore the systems to a past ideal but rather to restore and maintain the ecosystem services provided (e.g., Cairns 1989; Cairns and Heckman 1996).

A research initiatives subcommittee of LTER previously developed a strategic research plan to integrate science for society and the environment in which ecosystems services were emphasized (Collins et al. 2007). Ecosystem services were classified as *provisioning*, *regulating*, and *cultural* services. The roots of multiple-use management (select services) are deeply embedded in USDA Forest Service philosophy and policy, and in its public forests and grasslands management programs. This was true in the early history of the agency and was formally enacted in the Multiple-Use Sustained-Yield Act of 1960. This mandate provided the public with multiple benefits, including clean water, abundant fish and wildlife, a sustainable supply of wood and paper products, and ensured that there would be quality outdoor environments for recreation, wilderness and scenic rivers, supplies of energy and minerals; livestock foraging, and the development of human resources. A pilot program in the multiple-use concept was undertaken in a watershed context at the Coweeta Hydrologic Laboratory, beginning in 1962 (Hewlett and Douglass

1968) and it continues to provide a valuable example of blending forest benefits for more than 30 years (Swank 1998).

Over the years, the mix of forest uses and benefits evolved in response to changing public needs, new legislative mandates, improved scientific information, and advances in technology. To address changing views of land and natural resources, the Forest Service took a new direction in its research and management programs, adopting an operating philosophy of ecosystem management with the objective of using an ecological approach to incorporate an array of ecosystems services (Thomas 1996). The agency view of ecosystem management is to integrate ecological, economic, and social factors to maintain and enhance the quality of our environment to meet current and future needs.

A compendium of essays on ecosystem management was published in *Ecological Applications* in 1996 (Haeuber and Franklin 1996 and the following papers in that issue), which represented a wide range of view points on the topic. In 1991, a technical symposium was convened at the American Association for the Advancement of Science Annual Meeting that addressed the concept, philosophy, needs, and opportunities related to ecosystem management (Swank and Van Lear 1992). Research at two LTER sites (Hubbard Brook and Coweeta, including WS 7) were used to illustrate how watershed ecosystem analysis can be used to address forest environmental and management issues (Hornbeck and Swank 1992).

Ecosystem management has been interpreted and implemented in a variety of ways by different organizations and agencies. At Coweeta, beginning in 1992, we developed and initiated an ecosystem management demonstration project in the Wine Spring Creek basin in western North Carolina (Swank 1998). The 1130-ha watershed is covered with a mix of hardwood forest types, first- through third-order streams, and diverse flora and fauna. Most of the basin is made up of National Forest land. Participants in this project included an interdisciplinary team of over 55 scientists and managers in five research units in the Southern Research Station, National Forest Systems, and seven universities; state agencies; environmental and conservation groups; and the public. Desired future resource conditions and management prescriptions were determined through a series of workshops comprised of interested stakeholders. During the next decade, a large body of knowledge was derived from research for a wide range of land uses and ecosystem services. This data base was synthesized using an EMERGY-based environmental systems assessment (Odum 1996) of the services provided within the Wine Spring Creek basin (Tilley and Swank 2003). This method of assessment is a mechanism to express ecological and economic benefits with a common metric. The analysis included products and services such as recreation, biodiversity, water yield, timber production, biogeochemical cycles, and research information (Tilley and Swank 2003) and suggested that one of the largest benefits coming into the basin was the large number of scientists and managers involved in the ecosystem demonstrations project and the high value of their research. Christensen et al. (1996) also listed research as one of the services provided by ecosystems. The same is true as a result of the long-term research investment on WS 7; WS 7 has so far provided advanced degrees for more than 20 graduate students, and some 25 senior scientists conducted research on the watershed, publishing over 125 papers (figures 14.3 and 14.4).

Figure 14.3   Scientists gathered for a three-day Symposium held in Athens, Georgia, in October 1984 to commemorate 50 years of research at the Coweeta Hydrologic Laboratory. The goal of the symposium was to summarize and highlight the major contributions from Coweeta to the hydrologic and ecological understanding of southern Appalachian forested lands. The meeting was attended by over 75 individuals and resulted in a book with contributions from 49 authors/coauthors in 30 papers distributed across 8 topical sections of research. This photo was taken late in the symposium and did not include some of the participants. Keynote speakers included Eugene P. Odum, Jerry F. Franklin, and Hans M. Keller. (USDA Forest Service photo)

Ecosystem management is based on uncertain knowledge of dynamic ecosystems responding to uncertain predictions of the future (Holling 1996; Lawler et al. 2010). Therefore, ecosystem management must be adaptive (e.g., Walters 1986), changing in response to information provided by studies such as those on WS 7. Examples of this type of forward-looking management include the design of roads and culverts to handle greater rainfall, selection of tree species adaptable to predicted changes in climate and from disturbance regimes. In the Southeast, this might include species that use less water and those that are able to survive extended droughts. Species selection might also mean the choice of species that are resistant to diseases and insects, for example, gypsy moths. The phrase "engineering resistance" has been used in forestry to describe the genetic engineering of trees for resistance to disease. We suggest that the phrase might also be used more broadly to describe the management of forests to include structural attributes and species that may be more adapted to future environmental conditions. Similarly, the role of forests and the options for their management to mitigate greenhouse gas emissions has also been identified (Malmsheimer et al. 2008) and should be considered in any forward-looking research plan.

Figure 14.4 In November 2009, over 150 scientists gathered for three days in Dillard, Georgia, to commemorate 75 years of research at Coweeta. The focus of this symposium was on the benefits of the long-term research conducted at Coweeta in addressing fundamental hypotheses on hydrologic and ecological processes, as well as practical questions related to science-based land management. Many of the individuals in this picture also participated in the fiftieth anniversary celebration and authored papers in this volume. (USDA Forest Service photo)

As a site for research to provide guidelines for the adaptive ecosystem management of forested watersheds, WS 7 and other watersheds at places such as Fernow and Hubbard Brook provide an ecosystem service with a value far in excess of other goods and services. Meyer and Swank (1996) concluded that ecosystem management challenges ecologists to test theory in a real-world landscape laboratory. But ecosystem management must also respond to these research findings. As described by Stanford and Poole (1996), research and management policy must involve an iterative protocol—research based on current management strategies and management that incorporates new research findings. Using WS 7 and other watersheds at Coweeta, we will continue to develop hypotheses and test predictions for important ecosystem processes and system-level responses associated with both forest management and natural disturbances.

Literature Cited

Aber, J. D., K. J. Nadelhoffer, P. Steudler, and J. M. Melillo. 1989. Nitrogen saturation in northern forest ecosystems. *BioScience* 39: 378–386.
Arthur, M. A., H. D. Alexander, D. C. Dey, C. J. Schweitzer, and D. L. Loftis. 2012. Refining the oak-fire hypothesis for management of oak-dominated forests of the eastern United States. *Journal of Forestry* 110: 257–266.

Bender, E. A., T. J. Case, and M. E. Gilpin. 1984. Perturbation experiments in community ecology: theory and practice. *Ecology* 65: 1–13.

Cairns, J. 1989. Restoring damaged ecosystems: is predisturbance condition a viable option? *Environmental Professional* 11: 152–159.

Cairns, J., and J. R. Heckman. 1996. Restoration ecology: the state of an emerging field. *Restoration Ecology* 21: 167–189.

Chapin, F. S., and A. M. Starfield. 1997. Time lags and novel ecosystems in response to transient climatic change in arctic Alaska. *Climatic Change* 35: 449–461.

Christensen, N. L., A. M. Bartuska, J. H. Brown, S. Carpenter, C. D'Antonio, R. Francis, J. F. Franklin, J. A. MacMahon, R. F. Noss, D. J. Parsons, C. H. Peterson, M. G. Turner, and R. G. Woodmansee. 1996. The report of the Ecological Society of America Committee on the Scientific Basis for Ecosystem Management. *Ecological Applications* 6: 665–691.

Clawson, M. 1979. Forests in the long sweep of American history. *Science* 204: 1168–1174.

Clements, F. E. 1916. Plant succession: an analysis of the development of vegetation. Carnegie Institute of Washington, Publication No. 242. Washington, DC.

Clements, F. E. 1936. Nature and structure of the climax. *Journal of Ecology* 24: 252–284.

Collins, S. L., S. M. Swinton, C. W. Anderson, T. Gragson, N. B. Grimm, M. Grove, A. K. Knapp, G. Kofinas, J. Magnuson, B. McDowell, J. Melack, L. Ogden, O. J. Reichman, G. P. Robertson, M. D. Smith, and A. Whitmer. 2007. Integrative Science for Society and Environment: A strategic research plan. Integrative Science for Society and Environment: A Strategic Research Initiative. LTER Network Office Publication #21. Albuquerque, New Mexico.

Dolloff, C. A., and J. R. Webster. 2000. Particulate organic contributions from forests to streams: debris isn't so bad. Pages 125–138 in *Riparian Management in Forests of the Continental Eastern United States*. E. S. Verry, J. W. Hornbeck, and C. A. Dolloff, editors. Lewis Publishers, Boca Raton, Florida

Douglass, J. E., and M. D. Hoover. 1988. History of Coweeta. Pages 17–31 in *Forest Hydrology and Ecology at Coweeta*. W. T. Swank and D. A. Crossley, Jr., editors. Springer-Verlag, New York, New York.

Elliott, K. J., and W. T. Swank. 1994. Changes in tree species diversity after successive clearcuts in the southern Appalachians. *Vegetatio* 115: 11–18.

Elliott, K. J., and W. T. Swank. 2008. Long-term changes in forest composition and diversity following early logging (1919-1923) and the decline of American chestnut (*Castanea dentata*). *Plant Ecology* 197: 155–172.

Elliott, K. J., and J. M. Vose. 2005. Effects of understory burning on shortleaf pine (*Pinus echinata* Mill.)/mixed-hardwood forests. *Journal of the Torrey Botanical Society* 132: 236–251.

Elliott, K. J., and J. M. Vose. 2011. The contribution of the Coweeta Hydrologic Laboratory to developing an understanding of long-term (1934–2008) changes in managed and unmanaged forests. *Forest Ecology and Management* 261: 900–910.

Ellison, A. M., M. S. Bank, B. D. Clinton, E. A. Colburn, K. Elliott, C. R. Ford, D. R. Foster, B. D. Kloeppel, J. D. Knoepp, G. M. Lovett, J. Mohan, D. A. Orwig, N. L. Rodenhouse, W. V. Sobczak, K. A. Stinson, J. K. Stone, C. M. Swan, J. Thompson, B. Von Holle, and J. R. Webster. 2005. Loss of foundation species: consequences for the structure and dynamics of forested ecosystems. *Frontiers in Ecology and the Environment* 3: 479–486.

Ely, D. T., and J. B. Wallace. 2010. Long-term functional group recovery of lotic macroinvertebrates from logging disturbance. *Canadian Journal of Fisheries and Aquatic Sciences* 67: 1126–1134.

Foster, D. H. 1999. *Thoreau's Country: Journey through a Transformed Landscape*. Harvard University Press, Cambridge, Massachusetts.

Fox, S., G. Kauffman, M.C. Koester, C. Van Sickle, J. fox, J. Hicks, T. Pierce, B. Power, and L. Ross. 2010. The Western North Carolina Report Card on forest Sustainability. USDA Forest Service, Southern Research Station, Asheville, North Carolina.

Gregory, S. V., K. L. Boyer, and A. M. Gurnell, editors. 2003. *The Ecology and Management of Wood in World Rivers*. American Fisheries Society, Bethesda, Maryland.

Grimm, V., and C. Wissel. 1997. Babel, or the ecological stability discussions: an inventory and analysis of terminology and a guide for avoiding confusion. *Oecologia* 109: 323–334.

Gunderson, L. H. 2000. Ecological resilience—in theory and application. *Annual Review of Ecology and Systematics* 31: 425–439.

Gurtz, M. E., and J. B. Wallace. 1984. Substrate-mediated response of stream invertebrates to disturbance. *Ecology* 65: 1556–1569.

Gurtz, M. E., J. R. Webster, and J. B. Wallace. 1980. Seston dynamics in southern Appalachian streams: effects of clear-cutting. *Canadian Journal of Fisheries and Aquatic Sciences* 37: 624–631.

Haeuber, R., and J. Franklin. 1996. Perspectives on ecosystem management. *Ecological Applications* 6: 692–693.

Hewlett, J. D. and J. E. Douglass. 1968. Blending forest uses. Research Paper. SE-37. US Department of Agriculture, Forest Service, Southeastern Forest Experiment Station, Asheville, North Carolina.

Hobbs, R. J., S. Arico, J. Aronson, J. S. Baron, P. Bridgewater, V. A. Cramer, P. R. Epstein, J. J. Ewel, C. A. Klink, A. E. Lugo, D. Norton, D. Ojima, D. M. Richardson, E. W. Sanderson, F. Valladares, M. Vila, R. Zamora, and M. Zobel. 2006. Novel ecosystems: theoretical and management aspects of the new ecological world order. *Global Ecology and Biogeography* 15: 1–7.

Holling, C. S. 1973. Resilience and stability of ecological systems. *Annual Review of Ecology and Systematics* 4: 1–24.

Holling, C. S. 1996. Surprise for science, resilience for ecosystems, and incentives for people. *Ecological Applications* 6: 733–735.

Hooper, D. U., F. S. Chapin, J. J. Ewel, A. Hector, P. Inchausti, S. Lavorel, J. H. Lawton, D. M. Lodge, M. Loreau, S. Naeem, B. Schmidt, H. Setala, A. J. Symstad, J. Vandermeer, and D. A. Wardle. 2005. Effects of biodiversity on ecosystem functioning: a consensus of current knowledge. *Ecological Monographs* 75: 3–36.

Hornbeck, J. W., and W. T. Swank. 1992 Watershed ecosystems analysis as a basis for multiple-use management of eastern forests. *Ecological Applications* 2: 238–247.

Johnson, E. A. 2002. Industrial destruction reversed at Copper Basin. *National Woodlands* 25(2): 10–13, 29.

Lawler, J. J., T. H. Tear, C. Pyke, M. R. Shaw, P. Gonzalez, P. Kareiva, L. Hansen, L. Hannah, K. Klausmeyer, A. Aldous, C. Bienz, and S. Pearsall. 2010. Resource management in a changing and uncertain climate. *Frontiers in Ecology and the Environment* 8: 35–43.

Levin, S. A., and J. Lubchenco. 2008. Resilience, robustness, and marine ecosystem-based management. *BioScience* 58: 27–32.

MacCleery, D. W. 1992. *American Forests: A History of Resiliency and Recovery*. Forest History Society, Durham, North Carolina.

Malmsheimer, R. W., P. Heffernan, S. Brink, D. Crandall, F. Deneke, C. Galik, E. Gee, J. A. Helms, N. McClure, M. Mortimore, S. Ruddell, M. Smith, and J. Stewart. 2008. Forest management solutions for mitigating climate change in the United States. *Journal of Forestry* 106: 115–171.

Mann, C. C. 2005. *1491: New Revelations of the Americas before Columbus*. Knopf, New York, New York.

Meyer, J. L., and W. T. Swank. 1996. Ecosystem management challenges ecologists. *Ecological Applications* 6: 738–740.

Monk, C. D., D. A. Crossley, S. T. Swank, R. L. Todd, J. B. Waide, and J. R. Webster. 1977. An overview of nutrient cycling research at Coweeta Hydrologic Laboratory. Pages 505–526 in *Watershed Research in Eastern North America*. D. L. Correll, editor. Smithsonian, Washington, DC.

Nowak, J., Asaro, C., Klepzig, K., and Billings, R. 2008. The southern pine beetle prevention initiative: working for healthier forests. *Journal of Forestry* 106: 261–267.

Odum, H.T. 1966. *Environmental Accounting: EMERGY and Environmental Decision Making*. Wiley, New York, New York.

Palumbi, S. R., K. L. Mcleod, and D. Grunbaum. 2008. Ecosystems in action: lessons from marine ecology about recovery, resistance, and reversibility. *BioScience* 58: 33–42.

Richter, D. D., and D. Markewitz. 2001. *Understanding Soil Change: Soil Sustainability over Millennia, Centuries, and Decades*. Cambridge University Press, Cambridge, England.

Robinson, V. L., and E. L. Fisher. 1982. High-lead yarding costs in the Southern Appalachians. *Southern Journal of Applied Forestry* 6: 172–176.

Seastedt, T. R., R. J. Hobbs, and K. N. Suding. 2008. Management of novel ecosystems: are novel approaches required? *Frontiers in Ecology and the Environment* 6: 547–553.

Stanford, J. A., and G. C. Poole. 1996. A protocol for ecosystem management. *Ecological Applications* 6: 741–744.

Stone, M. K., and J. B. Wallace. 1998. Long-term recovery of a mountain stream from clear-cut logging: the effects of forest succession on benthic invertebrate community structure. *Freshwater Biology* 39: 151–169.

Stout, B. M., E. F. Benfield, and J. R. Webster. 1993. Effects of a forest disturbance on shredder production in southern Appalachian headwater streams. *Freshwater Biology* 29: 59–69.

Swank, W.T. 1998. Multiple use forest management in a catchment context. Pages 27–37 in Proceedings of an International Conference. Multiple Land Use and catchment management. M. Cresser and K. Pugh, editors. The Macaulay Land Use Research Institute, Aberdeen, Scotland.

Swank, W. T. 1984. Atmospheric contributions to forest nutrient cycling. *Water Resources Bulletin* 20: 313–321.

Swank, W. T., and D. H. Van Lear. 1992. Multiple-use management: ecosystems perspectives of multiple-use management. *Ecological Applications* 2: 219–220.

Swank, W. T., and J. M. Vose. 1997. Long-term nitrogen dynamics of Coweeta forested watersheds in the southeastern United States of America. *Global Biogeochemical Cycles* 11: 657–671.

Swift, L. W. 1988. Forest access roads: design, maintenance, and soil loss. Pages 313–324 in *Forest Hydrology and Ecology at Coweeta*. W. T. Swank and D. A. Crossley, Jr., editors. Springer-Verlag, New York, New York.

Thomas, J. W. 1996. Forest Service perspective on ecosystems management. *Ecological Applications* 6: 703–705.

Thoreau, H. D. 1860. The succession of forest trees. Read to the Middlesex Agricultural Society, Concorde, Massachusetts, USA, September 1860. Reprinted in 1863 in *Excursions*, pages 135–160. Ticknor and Fields, Boston, Massachusetts.

Tilley, D. R., and W. T. Swank. 2003. EMERGY-based environmental systems assessment of a multi-purpose temperate mixed-forest watershed of the southern Appalachian Mountains, USA. *Journal of Environmental Management* 69: 213–227.

Vose, J. M., and P. V. Bolstad. 2007. Biotic and abiotic factors regulating forest floor CO2 flux across a range of forest age classes in the southern Appalachians. *Pedobiologia* 50: 577–587.

Waide, J. B. 1988. Forest ecosystem stability: revision of the resistance-resilience model in relation to observable macroscopic properties of ecosystems. Pages 383–405 in *Forest Hydrology and Ecology at Coweeta*. W. T. Swank and D. A. Crossley, Jr., editors. Springer-Verlag, New York, New York.

Waide, J. B., and W. T. Swank. 1977. Simulation of potential effects of forest utilization on the nitrogen cycle in different southeastern ecosystems. Pages 767–789 in *Watershed Research in Eastern North America*. D. L. Correll, editor. Smithsonian, Washington, DC.

Waide, J. B., and J. R. Webster. 1976. Engineering systems analysis: applicability to ecosystems. Pages 329–371 in *Systems Analysis and Simulation in Ecology*. B. C. Patten, editor. Academic Press, New York, New York.

Walters, C. J. 1986. *Adaptive Management of Renewable Resources*. McGraw Hill, New York, New York.

Webster, J. R., S. W. Golladay, E. F. Benfield, J. L. Meyer, W. T. Swank, and J. B. Wallace. 1992. Catchment disturbance and stream response: an overview of stream research at Coweeta Hydrologic Laboratory. Pages 231–253 in *River Conservation and Management*. P. J. Boon, P. Calow, and G. E. Petts, editors. Wiley, Chichester, England.

Webster, J. R., M. E. Gurtz, J. J. Hains, J. L. Meyer, W. T. Swank, J. B. Waide, and J. B. Wallace. 1983. Stability of stream ecosystems. Pages 355–395 in *Stream Ecology*. J. R. Barnes and G. W. Minshall, editors. Plenum Press, New York, New York.

Webster, J. R., and B. C. Patten. 1979. Effects of watershed perturbation on stream potassium and calcium dynamics. *Ecological Monographs* 49: 51–72.

Webster, J. R., J. B. Waide, and B. C. Patten. 1975. Nutrient recycling and the stability of ecosystems. Pages 1–27 in *Mineral Cycling in Southeastern Ecosystems*. F. G. Howell, J. B. Gentry, and M. H. Smith, editors. National Technical Information Service, Springfield, Virginia.

Williams, M. 1989. *Americans and Their Forests: A Historical Geography*. Cambridge University Press, Cambridge, England.

Worrall, F., W. T. Swank, and T. P. Burt. 2003. Changes in stream nitrate concentrations due to land management practices, ecological succession, and climate: developing a systems approach to integrated catchment response. *Water Resources Research* 39. doi: 10.1029/2000WR000130.

Yarnell, S. L. 1998. The southern Appalachians: a history of the landscape. Gen Tech. Rep. SRS-18. Asheville, NC: U.S. Department of Agriculture, Forest Service, Southern Research Station.

Index